MATH ACTIVITIES FOR CHILDREN

A Diagnostic and Developmental Approach

MATH ACTIVITIES FOR CHILDREN

A Diagnostic and Developmental Approach

RICHARD W. COPELAND

Florida Atlantic University

Charles E. Merrill Publishing Company
A Bell & Howell Company
Columbus   Toronto   London   Sydney

Published by
Charles E. Merrill Publishing Company
A Bell & Howell Company
Columbus, Ohio 43216

This book was set in Optima and Souvenir
The production editor was Ann Mirels
The cover was prepared by Will Chenoweth

PHOTO CREDITS: Photos on pages 31, 60, and 138 from Richard W. Copeland, *How Children Learn Mathematics*, 3rd ed. (New York: Macmillan Publishing Co., Inc., 1979). Used with permission.

Photos on pages 34, 37, 42, 51, 66, 73, 87, 98, 129, and 153 from Richard W. Copeland, *Mathematics and the Elementary Teacher*, 3rd ed. (Philadelphia: W. B. Saunders Company, 1976). Used with permission.

Photos on pages 16, 24, 28, 47, 69, 85, 103, 104, 115, 126, 135 (and cover), 159, 166 (top and bottom), 172, 176, 193, 196, and 205 courtesy of the author.

Library of Congress Catalog Card Number: 78-60325
International Standard Book Number: 0-675-08316-8

1 2 3 4 5 6 7 8 9 10 / 85 84 83 82 81 80 79

Printed in the United States of America

Preface

Math Activities for Children is the outgrowth of a study of the research of Jean Piaget. The book provides a wide range of activities for teachers to use in the development of children's understanding of mathematics. Included are activities from three of the most recent books by Piaget on the subjects of memory, knowing versus performing, and chance and probability. The new ideas on which the activities are based prompt thought-provoking questions, even for those already familiar with Piaget's theories.

The reader may find the activities in this book almost shocking, inasmuch as they clearly reveal that children think and learn in ways that are very different from what has been commonly presumed. Many current teaching practices have yet to draw upon the research of Piaget. It is the author's hope that these activities will help to make a beginning.

Many of the photographs are from two textbooks the author has previously written: *How Children Learn Mathematics,* third edition, and *Mathematics and the Elementary Teacher,* third edition. For the reader interested in viewing a demonstration of the interview process described in *Math Activities for Children,* a 30-minute black and white film is available from the author. Featured are children 5 through 7 years of age.

To improve the process of teaching children mathematics, there is still much to learn about how children think. It is hoped this book provides some answers.

Richard W. Copeland
Boca Raton, Florida

v

Contents

NUMBER 39

SPACE ORIENTATION 93

MEASUREMENT 117

Length

Weight

Area and Volume

Time

KNOWING VERSUS PERFORMING 163

MATHEMATICAL MEMORY   179

CHANCE AND PROBABILITY   189

Introduction

PURPOSE

This book is a study of logical and mathematical thinking in children. It consists of activities that determine the quality or stage of a child's thinking. Children's performance in these activities can serve as a basis for planning an appropriate instructional program in mathematics.

Teachers have asked for a book that (1) incorporates the ideas of Jean Piaget, (2) presents these ideas in a form not difficult to understand, (3) considers their implications for the classroom, and (4) provides an easy-to-follow format. This book is an attempt to meet these objectives.

ORGANIZATION

The 62 activities in this book are grouped in seven basic areas: (1) logical classification, (2) number, (3) space orientation, (4) measurement, (5) knowing versus performing, (6) mathematical memory, and (7) chance and probability. The last three represent the most recent works of Piaget. The activities are sequenced (from the simpler to the more complex) for the intellectual development of children. The chart on pages 2–3 keys each activity to the appropriate age levels. This chart is also printed on the back cover for the reader's convenience.

SUGGESTED ACTIVITIES FOR EACH AGE

Ages	Logical classification	Number	Space orientation	Measurement	Knowing versus performing	Mathematical memory	Chance and probability
4	1–1, 1–2, 1–3, 1–7, 1–8	2–1, 2–2, 2–3, 2–4, 2–5, 2–6, 2–13, 2–15	3–1, 3–2, 3–4	4–4, 4–12		6–1	7–1
5	1–1, 1–2, 1–3, 1–4, 1–5, 1–7, 1–8	2–1, 2–2, 2–3, 2–4, 2–5, 2–6, 2–7, 2–8, 2–9, 2–10, 2–13, 2–14	3–1, 3–2, 3–4	4–1, 4–2, 4–4, 4–8, 4–12, 4–13, 4–14	5–1, 5–2, 5–3, 5–4	6–1, 6–2	7–1, 7–2, 7–3, 7–4
6	1–1, 1–2, 1–3, 1–4, 1–5, 1–6, 1–7, 1–8	2–1, 2–2, 2–3, 2–4, 2–5, 2–6, 2–7, 2–8, 2–9, 2–10, 2–11, 2–13, 2–14	3–1, 3–2, 3–3, 3–4, 3–5, 3–6, 3–7, 3–8	4–1, 4–2, 4–3, 4–4, 4–5, 4–6, 4–7, 4–8, 4–9, 4–11, 4–12, 4–13, 4–14, 4–15, 4–16	5–1, 5–2, 5–3, 5–4	6–1, 6–2, 6–3	7–1, 7–2, 7–3, 7–4, 7–5, 7–6
7	1–1, 1–2, 1–3, 1–4, 1–5, 1–6, 1–7, 1–8	2–1, 2–2, 2–3, 2–4, 2–5, 2–6, 2–7, 2–8, 2–9, 2–10, 2–11, 2–12, 2–13, 2–14, 2–15	3–1, 3–2, 3–3, 3–4, 3–5, 3–6, 3–7, 3–8	4–1, 4–2, 4–3, 4–4, 4–5, 4–6, 4–7, 4–8, 4–9, 4–10, 4–11, 4–12, 4–13, 4–14, 4–15, 4–16	5–1, 5–2, 5–3, 5–4	6–1, 6–2, 6–3	7–1, 7–2, 7–3, 7–4, 7–5, 7–6

8	1-1, 1-2, 1-3, 1-4, 1-5, 1-6, 1-7, 1-8	2-7, 2-8, 2-9, 2-10, 2-11, 2-12, 2-14, 2-15	3-3, 3-5, 3-6, 3-7, 3-8	4-3, 4-4, 4-5, 4-6, 4-7, 4-8, 4-9, 4-10, 4-11, 4-12, 4-13, 4-15, 4-16	5-1, 5-2, 5-3, 5-4	6-2, 6-3	7-1, 7-2, 7-3, 7-4, 7-5, 7-6
9	1-1, 1-2, 1-3, 1-4, 1-5, 1-6, 1-7, 1-8	2-7, 2-11, 2-12, 2-14, 2-15	3-3, 3-5, 3-6, 3-7, 3-8	4-4, 4-5, 4-7, 4-9, 4-10, 4-11, 4-12, 4-13, 4-15, 4-16	5-2, 5-3, 5-4	6-2, 6-3	7-1, 7-2, 7-3, 7-4, 7-5, 7-6
10	1-3, 1-4, 1-5, 1-6, 1-7, 1-8	2-12, 2-14, 2-15	3-5, 3-6, 3-7, 3-8	4-7, 4-9, 4-10, 4-11, 4-15, 4-16	5-2, 5-3, 5-4		7-1, 7-2, 7-3, 7-4, 7-5, 7-6
11	1-3, 1-4	2-12, 2-13	3-5, 3-6, 3-7, 3-8	4-7, 4-9, 4-10, 4-11	5-2, 5-3, 5-4		7-1, 7-2, 7-3, 7-4, 7-5, 7-6
12		2-12, 2-15	3-5, 3-6, 3-7, 3-8	4-7, 4-9, 4-10, 4-11	5-2, 5-3, 5-4		7-1, 7-2, 7-3, 7-4, 7-5, 7-6

JEAN PIAGET

Jean Piaget is a Swiss psychologist now in his eighties who has explored in great detail how children's mental processes work. In so doing, he has found that there are definite stages of intellectual development through which children go. While his studies have focused primarily on mental development rather than on social or emotional development, he does recognize the importance of social and emotional growth in the learning process.

Piaget's research has yet to be implemented in the educational process. Being concerned with what he characterizes as "logico-mathematical" knowledge, he has done extensive writing on the subject. Some of the most important sources as far as this book is concerned are the following:

	Published in English	Recent publisher
Piaget, *The Child's Conception of Number*	1952	W. W. Norton & Company, Inc., New York
Piaget and Inhelder, *The Child's Conception of Space*	1956	W. W. Norton & Company, Inc., New York
Piaget, Inhelder, and Szeminska, *The Child's Conception of Geometry*	1960	Basic Books, Inc., New York
Piaget and Inhelder, *The Early Growth of Logic in the Child*	1964	W. W. Norton & Company, Inc., New York
Beth and Piaget, *Mathematical Epistemology and Psychology*	1966	D. Reidel Publishing Company, Dordrecht, Holland
Piaget and Inhelder, *The Psychology of the Child*	1969	Basic Books, Inc., New York
Piaget, *The Child's Conception of Time*	1969	Basic Books, Inc., New York
Piaget, *Genetic Epistemology*	1970	Columbia University Press, New York
Piaget, *Science of Education and the Psychology of the Child*	1970	Orion Press, New York
Piaget and Inhelder, *Memory and Intelligence*	1973	Basic Books, Inc., New York

	Published in English	Recent publisher
Piaget and Inhelder, *The Origin of the Idea of Chance in Children*	1975	W. W. Norton & Company, Inc., New York
Piaget, *The Grasp of Consciousness — Action and Concept in the Young Child*	1976	Harvard University Press, Cambridge, Mass.

THE DEVELOPMENT OF KNOWLEDGE

Jean Piaget as a genetic epistemologist is concerned with knowledge and the processes by which it is acquired. He describes three basic types of knowledge.

1. The first type is a result of a **sensorimotor** intelligence which includes such instinctive-type behavior as sucking, blinking, and kicking. This type of knowledge exists in lower forms of animal life in a highly specialized but rigid way, and may be genetically determined; for example, processes that bees use in communicating location of honey or that birds use in annual migration.

2. A second type of knowledge may be characterized as **physical knowledge.** It is a factual-type knowledge based on sensory data, such as what we see or hear. For example, we see that leaves are green and that dogs run; we hear dogs bark and birds sing. But we may not be able to generalize that only some birds sing or that all things that sing are not birds. This involves a logical thought structure characteristic of the next level of knowledge.

3. The third and highest form of cognitive function is knowledge characterized as **logico-mathematical.** This knowledge often conflicts with physical knowledge or that which our senses tell us, as will be seen in the conservation activities. Logico-mathematical knowledge develops out of an *interaction* of our physical experiences and our logical mental processes. This knowledge develops in stages and is largely an internal process. It is not an acquired knowledge in the sense so many think of—the familiar stimulus-re-

sponse sequence which may be used to teach physical knowledge.

The problem of "teaching" logico-mathematical knowledge, then, is placed in a whole new context. Johnny may not be able to learn certain ideas regardless of how he is stimulated until a certain point in time when *he* has the necessary mental structures. Telling him about the transitive property, that is, if $a = b$ and $b = c$ then $a = c$, can have no meaning until he has the necessary thought processes. Thus, the teacher becomes a person who (1) uses the verbal medium to ask questions rather than to give answers and (2) provides physical experiences with objects that may serve as a basis for making relevant abstractions.

Levels of Intellectual Development

Children do not have the same mental processes available to them as do adults. When do intellectual changes take place in children and what are the characteristics of these changes?

Piaget finds four basic levels of intellectual development in children along with a number of substages.[1] These are characterized by name and approximate ages as

0 to 2 years	sensorimotor
2 to 7 years	preoperational
7 to 11 or 12 years	concrete operational
11 or 12 years on	formal operational

The ages given are approximations; they vary from activity to activity as will be seen later. The levels refer to logical and mathematical thought processes. At each succeeding level, the child has new processes available. The **preoperational** level is prelogical or pre-mathematical in that the child cannot understand such basic mathematical concepts as conservation of number and addition. Each activity in this book points out responses characteristic of this preoperational level. Responses at this level are usually based on perceptional cues, or how things look or appear to the child rather than the result of appropriate logical processes.

At the next level, **concrete operational,** the child uses logical or reasoning processes as a basis for an answer. At this level, the child

---

[1]For a full discussion see Jean Piaget and Barbel Inhelder, *The Psychology of the Child* (New York: Basic Books, Inc., 1969).

still needs to work with objects in order to formulate reasoning processes; hence, the name *concrete operational.*

The next level, **formal operational,** involves being able to think directly in the abstract—being able to generalize rules or definitions. Some of the activities in this book require the ability to think at the formal operational level; for example, activities involving proportions and probability.

Stages of Development

In each activity, responses are described as Stage 1, Stage 2, and Stage 3. Stage 1 responses are at the preoperational level; these responses involve no understanding. Stage 2 is a transitional one between the preoperational and concrete operational. The child works by trial and error to get correct answers, but is not convinced and may revert back to the preoperational level if the perceptual cues are more distorted. For example, the child thinks the following two rows have the same number of objects:

but thinks the following do not:

At Stage 3, the concrete operational level, responses are immediate and correct. Equivalence is not destroyed by lengthening or shortening the rows of objects. Children at this level do not have to use a trial-and-error approach; counting, for example, to see. (The Stage 1 child may even count correctly and still say the bottom row has more.)

For the higher level problems, such as those involving map making or probability, there is a Stage 4 that corresponds to the formal operational level. Stage 4 children are able to think at the abstract or theoretical level to formulate rules and to use proportions.

Factors Affecting Intellectual Development

While children do go through the four levels of intellectual development just described, the teacher should not be fatalistic. Piaget lists two factors, other than physiological and neurological, which affect intellectual development.

One of these factors is **social transmission.** Children do need to talk to their peers and to adults, to be asked questions and allowed to react to questions. They should not be "told" or "explained to," since this does not necessarily convey understanding. Children must be allowed to structure rationales for problem situations.

The other necessary factor is **physical experience.** The child must explore the objects around him. The concept of addition should grow out of combining and separating sets of objects. The idea of a square should grow out of handling, drawing, and describing one.

In conducting the activities for diagnostic and developmental purposes—if the child is Stage 1, allow several weeks of developmental time before considering the task again. If Stage 2, provide a variety of activities related to the same concept over a period of several weeks. If Stage 3, the child is, of course, ready for the related instructional process.

FUNCTIONS OF THE TEACHER

The preceding section makes certain teaching functions obvious. For example, one important function of the teacher is to *provide an environment that allows children to physically explore the objects around them.* From actions performed on objects, such as counting a row of buttons in both directions, a child discovers that the number is the same. The child may also find that 3 + 2 is the same as 2 + 3. In geometry, the child needs to physically explore shapes as a basis for the necessary space relation abstractions. Just looking at them and being told what they are is not enough.

Another related important function of the teacher is to *ask children questions rather than giving them answers or explanations.* Children must be able to structure mathematical or logical processes for themselves in order for explanations to be meaningful.

A third important function is to *allow children to interact with others* in exploring problem or concept situations.

A fourth important function is to *master an interview technique in order to effectively diagnose children's stages of logico-mathematical development.* Otherwise, children's stages of readiness will go undetermined, and they will not be placed where they should be in the math program. Students should take turns interviewing children or each other as the class observes, and then the interview procedure should be discussed. Most of the photographs in this book involve such a procedure.

MATH CONTENT AREAS

The seven content areas (Chapters 1–7) into which the 62 activities have been divided may be described briefly as follows:

Logical Classification

There is a prelogical period in children's development which corresponds to a prenumerical period. The construction of the idea of number in the mind of the child develops in close connection with the development of the ability to understand the logic of classification. Some classification activities should precede number activities. (See the chart on pages 2–3 [repeated on back cover] for keying activities to appropriate ages.)

Number

To understand number, children must understand the concept of conservation or invariance. They must also understand the concept of ordering (seriation) and that of class inclusion. In Chapter 2 are activities relating to each of these concepts.

Space Orientation

Children's concepts of space, or their geometrical concepts, are often different from those of the adult, as the activities in Chapter 3 on space demonstrate. Several types of geometrical concepts are considered including topological, Euclidean, and projective. The Euclidean concepts of horizontal and vertical, for example, are not abstracted until around 9 years of age. Yet these concepts are necessary for organizing spatial thinking for such tasks as understanding directions or making a map.

Spatial exploration of such ideas as sets of objects and number lines also serves as a basis for abstracting number ideas.

Measurement

Chapter 4 on measurement of length, area, volume, time, and motion is possibly the most interesting of all. The misconceptions of children are glaring, yet little recognition has been given to these misconceptions in teaching. The approach to teaching children "how to tell time," for example, has been very superficial, based almost entirely on a perceptual level. There was a temptation to place the chapter on measurement first for these reasons. But, developmentally, children are not fully operational with respect to some measurement concepts until 11 or 12 years of age.

Knowing versus Performing Mathematics

Addition as measured by "performance," "competency," or "behavioral objectives" is often described as "Given any two single-digit addends (such as 5 + 3), the child will name (write) the sum (8)."

But *knowing how* to give the answer in an addition problem is not the same as *knowing what* addition is. Being conscious of or understanding a logicomathematical process may lag by six years being able to "perform" the necessary operation correctly. This, of course, has extremely important implications from a psychological and an educational standpoint.

Mathematical Memory

Memory is often thought of as the storing of what is seen or heard and is assumed to be a true copy of reality. A consequent method of teaching is to "show" and have the learner "store" what is shown. But do children remember what they "see" or are shown, such as "This is a triangle" or "3 + 2 = 5"? Is memory perceptual in character as commonly supposed? Or does memory do its own thing based on the intellectual processes available to the child?

Piaget has found that a certain developmental level must be reached for children to remember what they have seen if a logical mathematical process such as ordering is involved. Thus memory is *not* a copy of what is seen. It is, instead, what the mind is able to reconstruct of that to which it has been exposed.

Chance and Probability

Many of the questions little children ask have no rational explanation, even though they think there is a hidden cause. For example, Why is this stick longer than that one? Why does Lake Geneva not go all the way to Berne? Why do your ears stick out? These famous "Why" questions demand a reason where a reason exists, but also where it does not—where the phenomenon happens by chance but the child sees a hidden cause.

Piaget's book, *The Origin of the Idea of Chance in Children* (1975), sheds much light on children's concepts of chance and probability. Chance as a cause of events is not even realized by young children. The formulation of the mathematics of probability is not fully developed in children until they are 12 to 15 years of age, but there is progress along these lines in the elementary school years, and there are appropriate activities for children then.

1
Logical Classification

There would seem to be two critical periods in the intellectual life of the child: the age 7–8 accompanied by a decline of egocentrism and the first appearance of the desire for verification or logical justification; and the age 11–12 when formal (deductive) thought first comes into being.[1]

The development of number does not occur earlier than that of classes (classifactory structures) or of . . . transitive relations [at 7 to 8 years of age on average].[2]

This parallelism between the evolution of number, classes and seriation [ability to order], is thus a first piece of evidence in favour of their interdependence as against the view that there is an initial autonomy of number.[3]

---

[1]Jean Piaget, *Judgment and Reasoning in the Child* (New York: Humanities Press, Inc., 1928), p. 74.
[2]Evert W. Beth and Jean Piaget, *Mathematical Epistemology and Psychology* (Dordrecht, Holland: D. Reidel Publishing Co., 1966), p. 259.
[3]Ibid., p. 261.

The activities in this chapter on logical classification are as follows:

1-1. Additive Classification

1-2. The Process of Foresight and Hindsight

1-3. The Relations of "Similar" and "Belonging to" in Classification

1-4. The Relations of "Some" and "All" in Class Inclusion

1-5. Hierarchical Classification

1-6. The Null Class or Empty Set

1-7. Multiplicative Classification (Intersection)

1-8. Simple Multiplication

ADDITIVE CLASSIFICATION

Purpose of Activity

The purpose of this activity is to explore children's ability to place objects in categories based on the likenesses or differences of the objects.

Materials Needed

Circles, squares, and triangles in two sizes, each size in two colors.[4] Shapes may be made of wood, plastic, or cardboard.

Procedure

Mix up the objects and place them on a table before the child. Ask him if he can put them in groups so that each group is alike in some way. If he can group by one criterion, such as shape, then mix up the objects again and ask if he can group another way (by size or by color).

Levels of Performance

Stage 1. This stage (Graphic Collections), ages 2 to 5, is characterized by an inability to group the objects by their properties of color, shape, or size. The child may respond to the directions by putting two shapes together, such as a small blue square and a small red square, and then switch his criterion to color and add a red triangle. He might then decide that a square placed below the

---

[4]Screens of gray have been used in conjunction with certain of the illustrations in the text to denote various colors. For example: 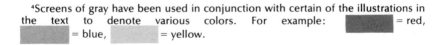 = red, = blue, = yellow.

triangle would make a house with a roof or other pleasing pattern. Even with repeated reminders to "put them together so they are alike," he is unable to continue to use the same criterion.

Stage 2.   At this stage (Nongraphic Collections), ages 5 to 7 or 8, the child is no longer tied to perception and is able to consider the objects as similar or dissimilar based on their properties of color, shape, or size. He proceeds to group them in one way, such as by shape, using a trial-and-error approach. He gradually develops a successful pattern. Progress may be aided by the use of boxes or sheets of paper and instructions to "put all that are alike on one sheet of paper," or "in one box."

Thus, the child at Stage 2 may be able to group the objects in only one way or two ways, and he uses a trial-and-error procedure.

Stage 3.   At Stage 3, from 6 to 9 years of age, the child's thinking is flexible. His thoughts are not tied to one configuration such as color. He can make the three classifications of color, shape, and size. He considers the problem and formulates a plan before he begins to work with the objects. He does not use a trial-and-error approach. Approximately one-third of children 6 to 7 years of age and two-thirds of children aged 8 to 9 will be in Stage 3 for this task.

Teaching Implications

Classification involves a study of relationships—how things are alike or different. Some of the first classifications by children are "Daddy," "doggy," and so on.

Logical classification is one of the most basic intellectual activities. It should precede work with number. Even the earliest programs of social studies, science, mathematics, reading, and spelling require the use of classification as a way of understanding our environment—naming and relating things.

Such available materials as buttons can be classified by color, shape, size, number of holes, and other properties. One of the properties of objects is their number, so the study of number can grow out of these activities.

Children should have many experiences in sorting common classroom materials, working with concrete shapes and sizes and colors, and discussing all sorts of relationships. Commercial materials such as logic blocks may be used, or materials such as those shown in the photograph on the following page.

It may be worthwhile to read the next activity on foresight and hindsight, so that it may be used in conjunction with this classification activity.

THE PROCESS OF FORESIGHT AND HINDSIGHT

Purpose of Activity

Foresight is characterized by the ability to consider and plan a method of action in problem solving. It involves anticipating results and mentally visualizing what would happen if a different plan were followed.

Hindsight is the process of reviewing past actions in the light of new information, as when additional items are introduced in a classification problem. Ability to change the criteria in view of new evidence is a matter of flexibility of hindsight.

Procedure

The previous activity, Additive Classification, may be used to study the child's use of foresight and hindsight.

Levels of Performance

Stage 1. The child at Stage 1, ages 2 to 5 or 6, is limited to a step-by-step performance in using foresight. His plans formulate as he works and encompass only the portion of the problem within his immediate view. He is incapable of considering the entire group of objects simultaneously and, therefore, cannot plan for the entire group. For example, he begins to classify by shape and then arbitrarily switches to color.

The Stage 1 child constantly shifts criteria in using hindsight. He has no coordinated use of all the necessary data.

Stage 2. At Stage 2, ages 5 to 7 or 8, a lack of foresight is still the primary obstacle to successful classification of the objects. The child seems unable to first form a mental picture of possible classifications to see if they would "work" and from this mental experimentation choose a procedure. Instead, he proceeds immediately with a trial-and-error operation in which his mistakes become apparent to him only after he has made them.

A lack of hindsight at Stage 2 results in a reluctance on the part of the child to give up any criterion once it has been used, even when it was only partially or not at all successful.

Stage 3. Flexibility of foresight is usually demonstrated at Stage 3, ages 7 to 9, by a period of planning before physical manipulation of the objects. The child may mentally select a criterion for classification, apply it to the objects, and judge its workability. He then proceeds with the task, working from the blueprint he has mentally constructed.

Flexibility of hindsight becomes important when the plan proves faulty or when new items must be considered. The child must be willing and able to scrap his old set of plans for a new one, yet keep his first plan available as a guide to what *not* to do.

Teaching Implications

The ability to coordinate a number of ideas that are necessary to make the right decision or to determine the proper course of action is difficult enough for adults. The ability of children to use foresight and hindsight is even more limited. The number of ideas relating to a course of action should be small at first, probably one, then two, and so on. Begin with one idea, such as "It is raining—what shall I do?"; then add a second, "It is raining and cold—what shall I do?"

THE RELATIONS OF "SIMILAR" AND "BELONGING TO" IN CLASSIFICATION

Purpose of Activity

Some small children, when asked to "put together the things that go together," respond by placing a block beneath a triangle and commenting that the "house" goes with the "roof." Until a child can differentiate between the house *belonging* to the roof, and the house being *similar* to another house, he is unable to begin true classification.

Materials Needed

Pictures or cutouts of

1. Four animals (for example, cat, dog, horse, cow).
2. Four humans—child, baby, cowboy, woman.
3. Four utensils—fork, knife, cup, glass.
4. Four articles of furniture—chair, table, bed, stool.

Procedure

Place the pictures on the table before the child. Encourage him to examine and discuss them.

Ask the child to tell you what the objects are. After he has familiarized himself with them, place six sheets of paper on the table and ask the child to put things on each sheet of paper so they are alike in some way. (The Stage 1 child may place the cowboy with the horse, or the woman with the chair, confusing "belonging" with "the same" and should be reminded, "Are they the same kind of things?") When the objects have all been grouped, one sheet of paper at a time should be withdrawn, forcing the child to further combine his collections. Behavior at this point reveals the ability to make a hierarchical classification.

Levels of Performance

Stage 1. At Stage 1, ages 2 to 5 or 6, confusion of the terms "similar" and "belonging to" is apparent and the child may fail to achieve the first step, even with constant reminders to aid his thinking. However, he is often able, with guiding comments from the examiner, to assemble six groups of similar objects. This is possible because he is required to consider only one object at a time in a succession of comparative decisions, and because the act of grouping objects on sheets of paper tends to divert his attention from the spatial arrangements involved. But when one sheet is removed and he is faced with the problem of considering the similarities between *groups* of objects, the child reverts to use of the belonging relation and is unable to determine classes that are more general.

Stage 2. At this stage, ages 5 to 7 or 8, the child is partially successful in combining his subgroups. He may put the cowboy with the child and woman as living things but not know where to put the baby. He is unable to think about the total picture before him. If he should successfully combine all of the objects into only two classes (such as living and nonliving), it would be the result of a series of hit-or-miss decisions, and the result would provide little insight for him into the general problem of classification. Also, as he attempts to combine classes, he is likely to revert back to the relation of "belonging to" rather than "similar."

Stage 3. The Stage 3 child, from 7 to 9 years of age, considers the sixteen objects in an overall way. He can make classifications and combine classes to make larger classes (hierarchical classification). He may divide the total group into classes of living and nonliving.

Teaching Implications

As children discover new things in their environment, they are faced with the problem of where these things "fit." How are they like other things? How are they different? Younger children will use the relation of "belonging to" and, for example, put a toy with a child rather than with other toys because it "belongs to" the child.

Early classificatory behavior in children is dominated by the "belonging to" as contrasted to the "similar" relation. Since the elementary teacher may have some Stage 1, some Stage 2, and many transitional children to deal with, an awareness of their patterns of thinking will help the teacher understand the difficulties children have with problems requiring logical classification.

THE RELATIONS OF "SOME" AND "ALL" IN CLASS INCLUSION

Purpose of Activity

The relations "all" and "some" are basic to classification—the "all" of a complete collection and the "some" of the subclasses within the collection. This activity reveals the extent to which the child experiences difficulty with these relations.

Materials Needed

Cardboard cutouts of three red squares, two blue squares, and three blue circles.

Procedure

1. First ask the child to identify the colors and shapes. "What color is this?" (pointing to one of the figures). "What shape is that?"
2. Then, the question, "Are all the circles blue?" This question is much less difficult than the next one.
3. "Are all the blue ones circles? Why?"

Children under the ages of 8 or 9 are usually unable to understand the correct answer—"No"—(because there are blue squares) even when it is pointed out or explained to them. They do not yet have the logical structures to consider the classes of blues, circles, and squares and the correct relations of "all" or "some" involved.

Levels of Performance

Stage 1.   The Stage 1 child knows only what he "sees." He cannot mentally separate the circles as a class from the whole series. "All" can only mean to him the whole of the graphic collection.

Stage 2.   At this stage (age 5 to 7 or 8) of nongraphic collections, the child has difficulties of a different sort. The very fact that he *can* disassociate the squares as a class from the circles and the reds from the blues makes his task more difficult, for he is not yet able to set up classes based on the logic of inclusion.

Q. Are all the circles blue?
A. No, because there are blue squares.

This is a typical answer and seems to stem from a misinterpretation of the question, but careful examination and wording will show that the verbalization is not at fault. The child interprets the question as "Are all the circles *all* the blues?" because he does not yet have the logical structures necessary to solve the question, "Are all the circles *some* of the blues?" The circles, thought about in this way, become a subset of blues, but not until Stage 3 is such class inclusion operational.

Stage 3.

Q. Are all the circles blue?
A. Yes.
Q. And are all the blue ones circles?
A. No.
Q. Why?
A. Because there are blue squares.

This child, aged 8 or 9, can establish logical classes of "circles,"

"squares," and "blues," and determine the relation between them in terms of "all" and "some." He is able to consider the entire heterogeneous grouping as "all" of the shapes, and the circles as "some" of the shapes which are blue.

Teaching Implications

An awareness on the part of the teacher of the difficulties involved in a some/all relationship will explain many inconsistencies in the prelogical child's behavior. Before age 9, use of the word "some" is a vague and intuitive process, and "all" loses its simplicity the moment it becomes a part of a larger "all."

HIERARCHICAL CLASSIFICATION

Purpose of Activity

Logical classification is a process of ordering and differentiating between classes, of combining parts into a whole and separating a whole into definable parts. In order to accomplish this process, a reversibility of thought is required, and it is necessary for the child to simultaneously consider the whole and its parts. This activity is particularly helpful for identifying the problems of the child who can recognize classes (Stage 2) but who fails to understand the hierarchical relation between classes; that is, flowers include roses, and roses include yellow roses.

Materials Needed

A set of pictures which form a hierarchical classification as follows:

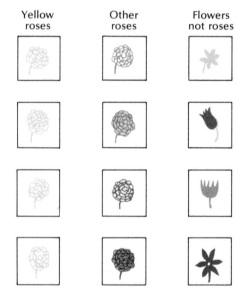

| Yellow roses | Other roses | Flowers not roses |

Procedure

Spread the pictures before the child and instruct him to "put them together so that they are alike." General questions as to color or type of object may guide his thinking, such as "If you made a bouquet out of all the flowers, would you use the yellow roses?" Then ask the child questions of the following four types:

1. Is a bunch made of all the yellow roses bigger or smaller than a bunch made of all the roses? Why?

2. Are there more roses or more flowers? Why?

3. If you take all of the roses, will there be any flowers left? Why?

4. If you take all of the flowers, will there be any roses left? Why?

Levels of Performance

Stage 1. The child in this age group (2 to 5) will be unable to construct the logical classes necessary to answer the preceding questions.

Stage 2. At Stage 2, age 5 to 7, the child can group the pictures qualitatively—that is, by similarities, such as putting the yellow roses together. Each grouping can be explained.

The following answers expose the basic problem of this stage:

Q. Is the bunch made of all the yellow roses bigger or smaller than the bunch made of all the other roses?
A. Bigger. I'll count them. Oh no, they are the same, four in each.
Q. Well, can I put this yellow rose with the other roses?
A. Yes. They're all roses.
Q. And do we have more roses or more yellow roses?
A. The same.

The child in considering the yellow roses, the roses which are not yellow, and *all* of the roses is unable to make the necessary classifications of "all" and "some" for quantitative comparisons.

Stage 3. At Stage 3, age 7 or 8, the child can answer the questions outlined in the procedure. He can consider the whole class (flowers) at the same time that he compares the subclasses (roses and yellow roses), and make the correct quantitative distinctions. The Stage 2 child, in contrast, when he looks at the subclasses (roses and yellow roses) cannot reverse his thought process back to the whole (flowers) and reason that logically there must be more flowers than roses or yellow roses since both of these classes are flowers.

Hierarchical classification of ducks, birds, and animals.

Teaching Implications

The problems experienced by the Stage 2 child's being taught addition facts may derive from the same source as do the problems with class inclusion—the inability to establish the proper relations between a whole and its parts. Can a child understand an addition problem such as $5 = 4 + 1$ if he cannot consider a bunch of five flowers in the form of four roses and a tulip? Asked if there are more roses or flowers, he responds, "Roses." Such children, once they establish the subclasses (roses and tulips), cannot reverse their thought process back to the whole (flowers) as far as quantitative comparisons are concerned.

Piaget used the terms **intention** and **extension** to describe going up or down in a hierarchical classification. Some children can go one way and not the other.

Intension, or going down the hierarchy, involves beginning with the whole—realizing that the flowers include roses (so there are more flowers than roses), and that the roses include yellow roses (so there are more roses than yellow roses).

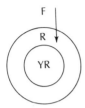

Extension begins with the set included in all the other sets and extends outward—realizing that yellow roses are included in the class of roses (so there are fewer yellow roses than roses), and roses are included in the flowers.

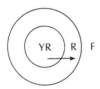

Similar activities can be conducted with pictures of children with subclasses of nine boys and two girls, for example. "Are there more boys or children? Why?" Or with a collection of pictures of ducks, birds, and animals—"If all the ducks died, would there be any birds left? Why?" And, "If all the birds died, would there by any ducks left? Why?"

THE NULL CLASS OR EMPTY SET

Purpose of Activity

The idea of the **empty** or **null set** may not be meaningful until the stage of formal operations (11 to 12 years of age), as this activity demonstrates. Most elementary school age children, being at the stage of concrete operations, may not be ready for it.

Materials Needed

1. Five triangular cards—three with pictures and two blank.
2. Five square cards—four with pictures and one blank.
3. Five circular cards—three with pictures and two blank.

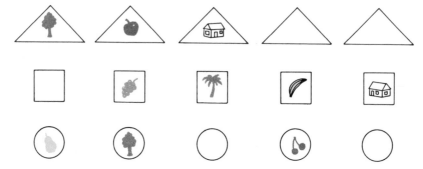

Procedure

Place the cards before the child with instructions to "put them together" or "classify" them in any way he chooses. Then ask the child to make another classification using only two groups.

Levels of Performance

Stage 1. The child at this stage (age 2 to 5 or 6) will be unable to make any distinct classifications.

Stage 2. Children aged 6 to 10 make the first classifications by varied criteria such as color of picture, type of picture, or shape of card. When asked to separate all the cards into only two piles, the problem of the blank cards becomes apparent. Some children may refuse to consider them a part of the activity since they have no picture. Others put the blank cards into groups of cards with pictures. Some children may yield to prompting and create a pile of cards with pictures and another pile of blank cards, but still insist that the blank set is not really a set at all.

Stage 3. The child aged 10 to 11 discovers the dichotomy of picture cards and blank cards and accepts it without worry or question. He is able to think beyond the objects represented on the cards and deal with the more abstract notion of a class with something in it and a class with nothing in it—the null class.

Teaching Implications

The need of most elementary-age children for a concrete base for thinking is often forgotten or minimized. It is important for the adult to realize how very different the thinking process is for these children and to carefully plan for the introduction of new ideas by use of concrete materials.

As far as the null class is concerned, the present practice of introducing it in the primary grades becomes a questionable one.

John, 9, unable to make a null-set classification, uses the blank cards as front yards for the houses.

MULTIPLICATIVE CLASSIFICATION (INTERSECTION)

Purpose of Activity

Preceding activities involved additive classification (a class of boys and a class of girls form a class of children). Two classes combine to form a new and higher class in a hierarchy of classification.

In **multiplicative classification,** each item will belong to two or more classes. Given a red square, blue square, red circle, and blue circle, they may be arranged vertically by color and horizontally by shape in a double classification. Square or rectangular grids (matrices) are convenient for studying such classifications.

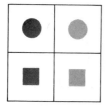

Multiple classification involves a more complex logical structure than does additive classification. From a psychological standpoint, the problems are often solved fairly early on a perceptual or graphic level because of the perceptual cues in the matrix.

Materials Needed

A set of large problem cards and smaller answer cards as shown on the following page.

Procedure

Place the first puzzle before the child and ask him to describe objects on it in terms of color, shape, size, number, and position of pictured object. Then give him the small item cards one at a time to see which fits the empty space on the large card. When the child has selected an item to complete the puzzle, ask him to justify his choice and, finally, to state whether another of the items would fit just as well or better and why.

1.

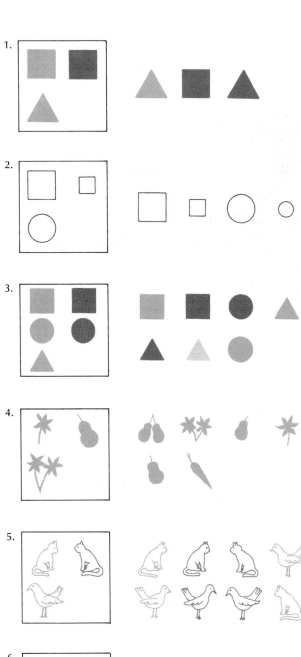

2.

3.

4.

5.

6.

Classifying rabbits as black, white, running, and seated.

Levels of Performance

Stage 1.   Children at this stage, 2 to 5 or 6 years of age, particularly in the 4-to-6 group, often choose the correct completing item. Surprisingly, they are more successful with the three-property items than with the two-property items. They are rarely able to justify their choice, however, and are easily moved to change it to another item. This testing situation seems to lend itself to graphic solution; the puzzle itself is a sort of symmetrical clue, and the three-property items simply provide more perceptual clues. The child cannot defend his choice with logic; he is often quite happy to switch to an aligning process and choose a duplicate of one of the items on the puzzle.

Stage 2.   During this period of transition, ages 5 to 7, the child often performs with less success than the Stage 1 child. Having given up the perceptual approach, he is committed to reasoning out the solution and does so with increasing ability as he progresses toward Stage 3. He correctly answers the two-property items well before he can cope with the more complex three-property items.

Stage 3.   This child, aged 7 or 8, solves each problem with little difficulty and is able to state the criteria for his choice in terms of similar or differing properties. He is also able to reject substitute choices and give the reason why they do not fit.

Teaching Implications

The two approaches to problem solving—the graphic or perceptually based and the operational or logically reasoned—are clearly demonstrated in the matrix test. The teacher becomes aware of the pattern of thought utilized by the child in solving these problems.

The idea of multiple classification involves an important idea in mathematics: the idea of multiplication and the Cartesian Product. Such activities as this involve readiness work for multiplication as a Cartesian Product.

SIMPLE MULTIPLICATION

Purpose of Activity

Simple multiplication may seem a less complex process than that of multiplicative classification, but it is not. Simple multiplication involves a mathematical idea known as **intersection.** Given two sets of objects, such as blue-eyed girls and brown-haired girls, the blue-eyed, brown-haired girls would be the intersection. This activity is designed to reveal the unsuspected difficulties of finding the common property or properties of two sets or classes.

Materials Needed

Card strip with pictures
of four red objects.

Card strip with pictures
of four green objects.

Card strip with pictures
of four blue objects.

Card strip with pictures
of four white objects.

Card strip with pictures
of four various colored cats.

Card strip with pictures
of four various colored leaves.

Card strip with pictures
of four various colored trees.

Procedure

Place before the child a strip of objects the same color; for example, the four red objects. At the end of this strip and perpendicular to it, place another strip of an object in different colors, for example, the cats.

36

Discuss each of the rows with the child. "How are all of the objects in this row alike? Why have they all been put together?"

Then ask the child to draw in the blank intersecting space the thing that fits or goes with each row, or to tell what would go in the blank space, or to pick from a collection of objects the one that would fit.

Levels of Performance

Stage 1. A completely perceptual solution is given at this stage with the child (aged 5 to 8) simply copying the closest item from one of the two rows. Additional rows may be added to create new intersections, but the child at Stage 1 responds with a consistent duplication of the nearest element.

Stage 2. This child, aged 5 to 8 or 9, is still unable to extract the two separate properties and combine them in a new form, but he uses more methods of approach than does the Stage 1 child. He may

a. select any one of the objects in either row for duplication;
b. add a completely different item because it can be associated—such as drawing a tree since both a pear and a leaf could fit with it; or

The 8-year-olds have difficulty finding "intersection" of sets.

    c. narrow his choice to two items, such as a green thing and a cat.

None of the Stage 2 solutions, however, successfully combine the two properties.

Stage 3. At age 8 or 9 the child analyzes the two rows and discovers the single determining property of each, then combines them in a single object. When all of the rows are presented, he is able to fill the point of intersection with a logical choice (solution) and to verbalize his procedure.

Teaching Implications

This idea, intersection, is important in mathematics and is often introduced to children before they are ready for it. This activity should provide good readiness information.

2
Number

What are the thought processes in the mind of the child that lead to an understanding of the logical concept of number?

There is a prelogical period in children's development corresponding to a prenumerical period, according to Piaget. The construction of the idea of number in the mind of the child develops in close connection with the development of the ability to understand the logic of class inclusion—flowers include roses, birds include ducks, and so on.

To understand number, the child must also understand the idea of ordering or seriation. In fact, psychologically speaking, number is a synthesis of classification and ordering. Ordering activities will be considered in this chapter on number. Classification activities were considered in the preceding chapter on logical classification.

Although addition is often taught now in the first grade, Piaget concludes, "We shall not here be concerned with investigating how the child learns addition and subtraction tables at school, learning which is frequently merely verbal.[1] . . . Additive composition comes late on the scene, in spite of appearances."[2]

Activities in this chapter provide insight into children's readiness to study number and the basic operations of addition, subtraction,

---

[1] Jean Piaget. *The Child's Conception of Number* (New York: W. W. Norton & Company, Inc., 1965), p. 161.
[2] Ibid., p. 198.

multiplication, and division. Properties related to these operations, such as the commutative, associative, and transitive, are also considered from the standpoint of readiness.

Activities included are as follows:

2-1. Ordering

2-2. Ordering and Seriation

2-3. Ordinal Number

2-4. Relating Ordinal Number to Cardinal Number

2-5. The Development of One-to-One Correspondence and Conservation of Quantity

2-6. Conservation of Number

2-7. Transitivity

2-8. Addition of Classes

2-9. Addition of Numbers

2-10. Addition and Subtraction (Making Quantities Equal)

2-11. The Commutative Property of Addition

2-12. The Associative Property of Addition

2-13. Multiplication of Numbers

2-14. Beginning Division

2-15. Fractions

2-16. Ratio and Proportion

ORDERING

Purpose of Activity

The purpose of this activity is to study children's understanding of the topological relation of **order**—a very important mathematical concept. To count, one must order the set of objects so that each object is counted once and only once. Shown a set of seven different colored beads arranged in a row, can the child (1) make another row with the same order of beads, (2) transpose the linear order to the same order in the form of a circle, (3) arrange the beads in reverse order, and (4) make a figure eight with the beads on a string of wire?

Materials Needed

1. Two matching sets of seven or nine different colored beads and some string or wire.

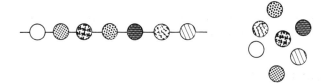

2. An alternative could be to put paper cutouts of clothing on a clothesline in a certain order and have children make a copy on another clothesline with a duplicate set of cutouts. Then have the child put the clothes on the clothesline in reverse order.

Procedure

Arrange one set of beads in a row on a wire. From a set containing beads of the same color, ask the child to make a row "like mine" on another wire. Then, with the model beads still in a row, ask the child if she can place hers on a wire in the form of a circle so that the order is the same. After that can she make a row that is the reverse of yours. You may need to help by starting the row for the child with the last bead and motioning backwards. Finally, using the model on

Michelle, 5, "orders" clothes on clothesline. She is unable to reverse order.

a piece of wire in the form of a figure eight, give the child another piece of wire also a figure eight and ask her to make a copy of yours.

Levels of Performance

Stage 1. Children between 2 and 3 are unable to make another row of beads in the same order. They may arrange two beads correctly in order but are unable to coordinate the whole sequence in a given order.

Stage 2. The 4- to 5-year-old can understand the notion of order and make a copy of a row of beads by constantly checking her row with the model. She is usually unable to transpose the linear order to a circular order and cannot make a row in reverse order.

Toward the end of Stage 2, the child may be able to reverse order, but it is a trial-and-error process. She will often miss after she gets to the center of the row, making the last half a copy of the model instead of in reverse order.

Stage 3. Between 6 and 7, the child achieves a rational and stable concept of order, solving the problems quickly and with ease. She sees the order between members of a series as a part of a unified whole. Having achieved reversibility of thought, children can now also reverse order and correctly consider the intertwining relation that exists in the figure-eight form.

Teaching Implications

Ordering is a fundamental mathematical activity. Ordering objects is important readiness work for ordering numbers. Reverse ordering is also important and necessary to understand numbers. That an operational understanding of order does not occur for many youngsters until 6 or 7 years of age should be an important consideration in lesson planning.

Children should have many opportunities to order, and of course they do. How do you set the table? Who sits here? Who sits next to him? Where does the fork go? The spoon? Who's first in line? Last? What goes on top? On the bottom? In the middle? Who is next to last? And so forth.

Many "pattern" games can be played for readiness and reinforcement. Using red and blue counters, who can tell what goes next?

R   B   R   B

R   B   B   R

R   R   B   R

R   B   B   B   R

Who can make up another pattern game?

Since number positions are not just assigned to objects in a row, other configurations should also be considered. Can the younger children reproduce such orders and discuss the characteristics of up, down, sideways, in the middle, and so forth? These are, of course, geometrical problems. Cardinal number is involved in terms of the correct number of objects and ordinal number in terms of describing the position of a particular object.

Ordering activities can be combined with shape familiarization. For example,

While the number of order possibilities for arranging five objects is not taught to young children, it is interesting that there are five factorial ways. This problem is solved as 5 × 4 × 3 × 2 × 1, or 120 different ways. Six people can be seated at a table in 6 × 5 × 4 × 3 × 2 × 1, or 720 ways. This is a mathematics called **permutations** (see Activity 7-6). Use fewer objects with young children. Two objects can be arranged in 2 ways, three objects or 3! is 3·2·1 or 6 ways, four objects is 4! or 4·3·2·1 or 24 ways.

ORDERING AND SERIATION

There is a very primitive ordering structure in children's thinking, just as primitive as the classification structure . . . the
structure of seriation.[3]

Purpose of Activity

The psychological structure in children's thinking necessary to understand the important mathematical concept of ordering Piaget
calls **seriation.** As early as 1 year of age, or when the child is still at
the sensorimotor level, she can order three objects such as bricks by
size if the size differences are easy to recognize, thus solving the
problem on a perceptual level. However, she is incapable of seriating as the number of objects becomes larger, or as the differences in
size become slight. What are the difficulties involved?

Activity 2-1 involved ordering a set of objects on which no relation
such as length had been imposed. In this activity, each object has a
unique length relation to each of the other objects.

Materials Needed

Ten sticks, graduated in size—the shortest about 5 centimeters in
length and each succeeding one longer by 0.5 centimeter.

Procedure

With the sticks mixed up, a 4-year-old is interviewed:

"Show me the smallest stick." She points to the right one.
"Now show me a stick a tiny bit bigger." She picks out a big one.
"Show me the biggest." She picks out a big one at random.
"Now try to find the smallest, then one a little bit bigger, then
another a little bit bigger." She picks sticks at random.

Then order the first 3 sticks so that they make a staircase and ask
the child to continue it. The child picks sticks of any length, but
arranges them so their tops do make a staircase outline. (Children at

---

[3]Jean Piaget, *Genetic Epistemology* (New York: Columbia University Press, 1970),
p. 28.

this level, under 4½ years of age, do not understand verbal descriptions of mathematical relations such as "more than," "less than," and "taller than.")

Levels of Performance

Stage 1. At the preoperational level, children cannot coordinate the series. They pick out a little stick and then a big one, another little one and another big one, or they pick sticks in groups of three—a little one, middle-sized one, and big one. There is no overall coordination.

Stage 2. The 5- to 6-year-old child has an "intuitive" idea of the series. She may construct the series partially or by trial and error construct the whole series correctly—trying first one stick and then another. That she still does not have a complete coordination of the necessary operations can be shown by giving her a few more sticks "that had been forgotten" and asking her to put them in their right place between the sticks already ordered. (The interviewer may place one stick himself as an example of what is desired.) The Stage

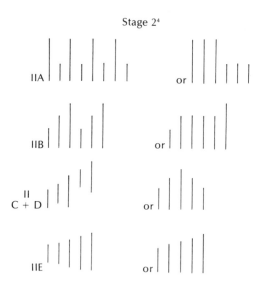

Stage 2[4]

---

[4]Jean Piaget and Barbel Inhelder, *Memory and Intelligence* (New York: Basic Books, Inc., 1973), p. 29.

2 child considers the series already built to be complete and feels no need to insert the additional sticks.

Stage 3.   From 6 to 7 years of age, the concrete operational level, children have a totally different and systematic way of solving the problem. They first find the smallest stick, then look through the collection for the next smallest, then for the next smallest, and so on until the whole series is built. These Stage 3 children are using the structure or logic of **reversibility** (reciprocity). They realize that each stick is both longer than the preceding one and shorter than the one to follow. They are coordinating the relation "longer than" and the relation "shorter than."

These children are also using the structure or logic of **transitivity.** They realize that if the third stick is longer than the second and the second is longer than the first, then the third must be longer than the first. Thus the whole series is constructed. Understanding the concept of transitivity can be investigated by hiding the first stick and asking the child how the length of the third stick compares with it. Preoperational children will say they do not know because they haven't "seen" them together. Their answers are based only on the perceptual means of "seeing" that one stick is longer.

Teaching Implications

To be successful in problems of seriation or ordering, it is important to keep in mind the necessary psychological structures of

1. reversibility of thought, or ability to order in two directions, such as forward and backward;

2. transitivity—if B is greater than A and C is greater than B, then C is also greater than A, thus coordinating a series of relations (around 7 years of age); and

3. the dual relations involved for any given element in determining its position—that it must be larger than the preceding element and yet smaller than the element to follow.

ORDINAL NUMBER

Purpose of Activity

If children are successful in seriating or ordering a set of sticks in order of length from shortest to longest, can they also make a double seriation; that is, arrange two sets of objects from shortest to longest and make a one-to-one correspondence between the two sets?

More important, can children pick out an object in one row and determine what object "belongs" to it in the other row if one row is spread out more? This involves making an **ordinal correspondence.** The child must find the position of a doll by counting from one end and then repeat this procedure in the other row.

Materials Needed

Paper cutouts or, if possible, the objects themselves:

1. Ten dolls, graduated in size, the last twice as big as the first.
2. Ten balls, graduated in size but not quite as much difference in size as the dolls.

Procedure

1. Show the child the dolls and the balls which are not arranged in any pattern. Ask her to pretend that the dolls want to play with their balls and to arrange the dolls so that it will be easy for each doll to find its ball.

49

2. When the child has ordered the dolls and balls by size, one row above the other, move the dolls closer together as the child watches. Then point to a doll and ask, "Which ball goes with this doll?"

3. Reverse the order of one of the rows (from largest to smallest for one row and from smallest to largest for the other). Then point to a doll and ask the child which ball goes with it.

4. Finally, with both the dolls and balls in no order, ask the child to find the right ball for a particular doll.

Levels of Performance

Stage 1. Children 4½ to 6 years of age arrange the set of dolls in a haphazard order. Asked to put the biggest first, then the one that's a little bit smaller until they come to the smallest, they change the order of the dolls but still do not arrange them correctly by size.

Asked what ball goes with the biggest doll, a 5-year-old points to the correct one. Asked what ball goes with the smallest doll, she points to the correct one. She is then asked to complete the row so that each doll has the right ball, but she is unable to do so.

Stage 2. Children from 5½ to 7 years of age are able to solve the problem of serial correspondence, but they use a perceptual method involving trial and error. They try one and then another until the serial correspondence between dolls and balls is established.

A child of 6½ years at Stage 2 is asked which ball goes with the biggest and smallest doll. She answers correctly, but when the interviewer points at the fifth largest doll and asks which ball goes with it, the child points at the seventh largest ball. Asked if there is any way we can be sure, this child says she will "put them like this." She arranges the dolls in order by size as follows: 10, 8, 9, then 10, 9, 7, 8, then 8, 7, 5, 6, and then 6, 5, 4, 3, 2, 1. Errors are corrected as she looks at the row and "sees" that doll 7 should not be between 8 and 9, for example.

The child then puts the balls opposite the dolls but makes an error of one position in each case, so that when she gets through there is one doll without a ball and one ball without a doll. Asked if there is the same number of dolls and balls, she says there are more balls. Asked to count, she finds 10 for each set and then realizes there are the same number of dolls and balls. She then corrects her error, arranging the two series correctly and making the correct one-to-one correspondence so that each doll has its ball.

At Stage 1, the child cannot make a serial correspondence. At Stage 2, the child can, but perceptual methods based on trial and error are used.

Stage 3.   At this stage, the child uses operational or intellectual procedures to solve the problem, but the differences in Stage 2 and Stage 3 methods are not easy to detect. At Stage 3, the correspondence made between dolls and balls is truly an ordinal or numerical one, as contrasted to the purely perceptual correspondence in Stage 2. The child described in Stage 2 has to have the number idea involved called to her attention, and she measures the dolls in pairs side by side before placing them in the series.

The child at Stage 3, around 7 years of age, considers all the dolls and balls at the same time rather than considering small groups or just pairs. Stage 3 children may not even feel a need to arrange the balls in a series. Some make a direct correspondence—picking out the biggest ball for the biggest doll, then the next largest ball for the next largest doll, and so on.

Other Stage 3 children immediately and correctly arrange the two series of dolls and balls by size, so that each ball is in front of the correct doll. However, if the row of dolls or balls is spread out,

Wendy, 6, Stage 3, can make an ordinal correspondence—can find the "right" stick for each doll.

there is not only the conservation of number problem, but the children must find some means other than perceptual to find what doll goes with what ball. Asked to find the ball that goes with a certain doll (pointing at the sixth doll, for example), the child at Stage 3 can use the idea of ordinal correspondence; that is, find out by counting that the doll is in the sixth position and then conclude she must count to the sixth position in the row of balls to find the right ball.

Teaching Implications

This very simple activity (for adults) is a very penetrating one as far as studying children's ideas of order and ordinal number.

The idea of ordinal number becomes operational in the child at approximately the same time as cardinal number. To Piaget, this is not surprising since he considers cardination to always involve ordination and vice-versa. To determine the number of objects in a set (cardinal number), the objects must be considered in some order if each is to be counted once and only once.

This activity could be varied in many ways, using double series of dolls to dresses, for example. This would be a more thorough approach to teaching children the idea of ordinal number than the simple exercises used in many first grades, such as having the children write the numerals corresponding to certain coaches of a train. In such tasks the problem is structured for the child.

RELATING ORDINAL NUMBER TO CARDINAL NUMBER

Purpose of Activity

The purpose is to determine how children relate the idea of ordinal number to the idea of cardinal number. To determine how many objects are in a set, the objects have to be ordered in some way so that each object is counted once and only once.

Materials Needed

Ten pieces of cardboard, cut such that each is one counting unit longer than the preceding one. If A is considered 1, then B is two times A or 1 + 1. Similarly, C is three A's, D is four A's, and so on.

A   B   C

Procedure

Having asked the child to order these 10 pieces of cardboard from shortest to longest, ask for the number of A's in B, in C, and so forth. Instead of using the letters, point to the first card and then to the second and third cards, asking "How many of these [A] make one of these [B]?" and then, "How many make one of these [C]?"

Does the child understand this pattern? Ask how many like A it will take to make one like this (pointing at the sixth card, for example). If the child wants to take the A card and measure off on the sixth card to find the answer, she does not see the relation of ordinal to cardinal number. However, if she counts from A to find that the card indicated is the sixth, and can then conclude that there must be six A's in it, she realizes the relation of ordinal to cardinal number. If the card is in sixth position (ordinal number), then its cardinal number value in terms of A or 1 must be six.

Levels of Performance

Stage 1. Children under 5 years of age are unable to order the cards by length beyond the first three or four. Asked how many like A does it take to make one like B, a child at this stage says that it takes two. To make a card like C she says that it takes four like A, then corrects herself saying three. But the next card, D, she says it takes five like A instead of four.

The interviewer then orders the cards for the child from shortest (A) to longest (K):

"How can we see that this one [pointing at D] is bigger than this one [C]?"

"By that much," pointing to difference in height of C and D.

"How many more are there each time?"

"One."

"Then how many cards like this one [A] can we make with that one [D]?"

"Five, no, two."

"How many?"

The child then counts the whole series A to K as 1, 2, 3, 4, 5, 6, 7, 8, 9, 10, but asked again how many like A make one like D she answers "five."[5]

Stage 1 children can count the number of cards and know the difference is one between two successive cards, or A is one. Yet they are unable to use *ordinal* position of a given card such as D and conclude that since it is in the fourth position, there must be four cards like A in it.

Stage 2. At around 5 years of age, children are able to arrange or order the cards from shortest to longest by a trial-and-error method. However, they are baffled when cards are picked at random or if the order of the testing from left to right is reversed by beginning at the right with the longest cards.

A 5-year-old constructed the series A, B, C, D, F, G, then inserted E, and by trial and error arranged the whole series correctly:

"If we cut this one [pointing at B], how many little ones like A can we make?"

"Three."

The interviewer puts A and B side by side and the child corrects her answer.

---

[5]Jean Piaget, *The Child's Conception of Number* (New York: W. W. Norton & Company, Inc., 1965), p. 136.

"No, two."

"And out of *D*?"

"Four."

The child finishes the series correctly, after which the interviewer asks, "And this one?" (pointing at *J* again).

"Nine."

"And this one?" (pointing at *H*).

"Ten."

"And this one?" (pointing at *G*).

"Eleven."

Stage 3.   From 6 to 7 years of age, the child has a complete understanding or an operational coordination of ordination and cardination. The interviewer can skip around in the series and the child answers correctly and immediately. If the card indicated is in the seventh position, the child knows that there are seven cards like *A* in it. She may count down—10, 9, 8, 7—instead of making the longer count from *A* or 1. If the cards are disarranged and the card *G* indicated, she orders the cards from *A* to *G* and counts to find seven.

Thus the series is no longer rigid at Stage 3, but is mobile or operational. Each card can be considered in relation to all the others.

Teaching Implications

Children must realize that the idea of number can be used to indicate position, such as sixth. This is the concept of ordinal number.

Number is also used to indicate "how many": How many objects are there in the set? There are six, for example. This is the concept of cardinal number. Cardinal number involves the use of ordinal number because the objects have to be ordered in some way to be sure that no objects are counted twice or are not counted at all.

Children need activities involving *both* ordinal and cardinal number. Most children's math books currently in use include many activities relating to cardinal number. Few offer a variety of activities relating to ordinal number. Fewer still offer activities relating cardinal number to ordinal number, such as Activity 2-4 just described. A variety of similar activities should be developed.

THE DEVELOPMENT OF ONE-TO-ONE CORRESPONDENCE AND CONSERVATION OF QUANTITY

Purpose of Activity

While **one-to-one correspondence** is the basis for counting to determine how many, how meaningful is it to children?

Materials Needed

1. Two containers of different shapes.
2. Beads.

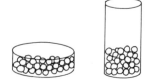

Procedure

The child is asked:

"Put a bead in your glass each time I put one in my glass."

When the beads in the tall container stack up higher, ask the child if each has the same number as the other, and if there is any way to be sure.

"What if we took them out and put them back one at a time in each glass?"

Levels of Performance

Stage 1. At 5 to 6 years of age, the idea of "one for you, one for me" is not sufficient to establish equality if the counters are placed in different-shaped containers. As the beads rise higher in the tall, slender glass, the child says it has more. Why? "Because it's higher." Thus, the beads are evaluated as if they were liquid in the containers and, not having the concept of conservation of quantity, the child relinquishes the idea of one-to-one correspondence for determining how many.

Stage 2. The child at this stage is transitional. She is swayed by first the one-to-one correspondence she has just performed and then by how the result looks. As perceptual distortions are increased (by using taller, thinner glasses, for example), she no longer thinks the amounts are the same. As Tis says, "You have got a lot. . . . I've not got so many, but a lot all the same."[6]

Stage 3. The child now does not have to reflect in order to be certain; she knows immediately the number is the same based on the one-to-one correspondence she has just carried out. She used the logic of one-to-one correspondence rather than perceptual cues as a basis for her answer.

Teaching Implications

This activity points out the extent to which children understand the idea of one-to-one correspondence as a basis for conservation of number. While one-to-one correspondence is a logical basis for determining equality, it is not so fixed in the child's mind that the concept will not be abandoned in the face of a more difficult one—conservation of quantity. As the beads pile up in the jars, the preoperational child abandons the idea of one-to-one correspondence and judges on the perceptual cues of height or width.

---

[6]Jean Piaget, *The Child's Conception of Number* (New York: W. W. Norton & Company, Inc., 1965), p. 32.

CONSERVATION OF NUMBER

Purpose of Activity

It is often assumed that if a child can say how many objects there are in a given collection, she understands what the number means. This activity demonstrates that counting, or one-to-one correspondence, is *not* the main criterion by which to judge the child's understanding of number. Does the child still think the number is the same when objects in a set are spread apart? Does she have the concept of **conservation of number?**

Materials Needed

1. Cardboard cutouts (or imitations) of seven flowers and seven vases.
2. Pennies and counters such as beans or buttons.

Procedure

1. With the seven vases in a row and a pile of flowers nearby, ask the child to get one flower for each vase and then to check her work by putting each flower in a vase (thus establishing a one-to-one correspondence).

   Then remove the flowers, put them in a bunch, and ask the child if there is still the same number of flowers and vases, or if there are more flowers or more vases.

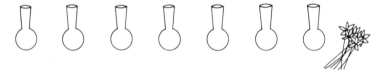

2. Vary the procedure using pennies in place of vases, telling the child she can buy one flower with one penny. Ask her to find out how many pennies she needs to buy all the flowers.

3. Use objects qualitatively the same, such as beans or pennies. Show one row of beans and ask the child to make another row with just as many beans as the first.

4. Show the child a collection of counters, eight or twelve, for example, arranged in a circle, square, or rectangular pattern and ask her to make another group with just as many in it. If she can do that, spread her group apart and ask if there is still the same number in hers as in yours.

Levels of Performance

Stage 1.   Children of 4 to 5 years of age often cannot make the one-to-one correspondence of flowers to vases. They make a row of flowers below the row of vases such that the end flowers line up with the end vases, but there are more or less flowers in the row than vases.

This, Piaget calls a **global evaluation.** The flowers are not thought of separately. If the ends line up, the child thinks the number is the same.

Stage 2.   Children of 5 to 6 or 7 years can make a one-to-one correspondence so that the row of vases is the same in number as the row of flowers. But if the flowers are put in a bunch, the child will say that there are more vases. If the flowers are spread out and the vases clustered together, she will say there are more flowers.

Such children, although they can make a one-to-one correspondence, do not establish a **lasting equivalence.** Perceptual factors (that is, the row of vases is longer so it must contain more) cause them to answer incorrectly.

Toward the end of Stage 2, as the child puts the flowers back in the vases and sees that the number is again the same, she realizes the conflict between what her senses tell her and what actually happens. Thus, by trial and error, she finds the correct answer.

Stage 3.   Usually between 6 and 7 years of age children have immediate and correct solutions to the problem. They realize the number does not change as the objects are spread out or bunched up. They are operational, can think logically, and realize what a number means. Perceptual factors, such as the length of the row of objects, are no longer the basis for the decision on how many.

Teaching Implications

Four variations of this activity were outlined in the procedure. Since it would seem easier to make a one-to-one or provoked correspondence for objects that go together (straws to bottles of milk, dolls to dresses, and so on), we began with flowers and vases. Piaget found, however, little difference in the age at which children could perform these different tasks, even though the first variation seems easiest.

The teacher might begin using sets of five objects and then seven or eight. Children in a transitional stage, Stage 2, may be successful with sets of five objects but unsuccessful with larger sets. As more objects are used or they are spread further apart, perceptual factors again hold sway producing wrong answers.

The last activity in the procedure (4) is useful since numbers are not always applied to objects in rows. Can the child make this transition?

When is a child ready for understanding number? Correct counting is not sufficient. A child may count correctly and say that each set has six objects; yet she may still say one set has more. The child must be aware of the invariance or conservation of number as a set of objects is rearranged.

Children should have practice in matching sets of objects to determine whether they are equivalent. They also need to learn the number names in order to count. But the teacher must realize that an important limitation may still be present: the children may not yet have the ability to conserve.

Piaget considers that the logical concept of conservation is a necessary condition for all reasoning or rational activity. Number is such an activity. The child must be working on a logical rather than on a perceptual level in order to understand number. Piaget concludes:

> Yet although one-one correspondence is obviously the tool used by the mind in comparing two sets, it is not adequate, in its original form or forms, to give true equivalence to the corresponding sets, i.e. to give each set the same cardinal value.[7]

---

[7]Jean Piaget, The Child's Conception of Number (New York: W. W. Norton & Company, Inc., 1965), p. 41.

TRANSITIVITY

Purpose of Activity

The idea of **transitivity** involves relating three or more elements in a certain way. For example, if B is taller than A and C is taller than B, then what is the relation of C to A? Logically, C is taller than A.

Expressed symbolically (with $B > A$, meaning B is greater than A), if

$B > A$

and

$C > B$

then

$C > A$

The relation "greater than" is a transitive relation. So is the "equals" relation. If $3 + 2 = 5$ and $5 = 4 + 1$, then $3 + 2 = 4 + 1$.

Transitivity is involved in many ways in solving mathematical problems.

1. If 6 is greater than 5 and 7 is greater than 6, then what is the relation of 7 to 5?
2. If a ruler is the same length as one tower and also the same length as another tower, what can be said about the heights of the towers?
3. In understanding time, do different clocks tell or measure the same time of an event?
4. If John lives farther from school than Mary and Bill lives farther than John, then who lives farther from school—Bill or Mary?

This activity helps to determine the extent to which children are able to use the important mathematical construct of transitivity.

Materials Needed

Four glasses corresponding in shape to those in the illustration, three containing the same amount of liquid.

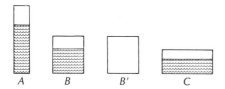

Procedure

Pour *A* into *B'* and ask the child if *B* and B' contain the same amount. If she does not think so, ask her to make them equal. After this has been done, pour *B'* back into *A*. Then repeat the procedure with *C* and *B'*, pouring *C* into *B'*, having the child agree *B* and *B'* are equal, and then pouring *B'* back into *C*. Then ask the child, "If you drank the liquid in *C* and I drank the liquid in *A*, would one of us drink more or would we drink the same?"

In the interview, point at the appropriate containers such as *A* and *B* saying, "Is this equal to that?" rather than using the letters.

Levels of Performance

Stage 1. At 5 to 6 years, children are preoperational for both transitivity and conservation of quantity (see also Activity 2-5). Their judgments are based on perceptual cues—one has more because it "looks" like more.

Stage 2. At 6 to 7 years, the amounts are estimated correctly, but no valid explanation can be given.

Stage 3. At 7 to 9 years, most children have grasped the principle of transitivity and can explain the principle involved: If *A* poured into *B'* is the same as *C* poured into *B'*, then *A* and *C* must be equal.

Teaching Implications

According to Piaget, the grasp of transitivity in this activity implies the grasp of conservation of quantity, since there must be a presupposition that the amount of liquid does not change during the pouring.

Some children grasp the conservation during a single pouring but do not understand the transitivity principle involved in the two related pourings.

Piaget also related this activity to memory by asking the children a week later to draw what they had seen the week before. At Stage 1, children "remembered" the most improbable data, such as pouring the liquid from A and C into B' at the same time. Memory cannot "recall" or copy what is "seen" if the idea involves a logical structure, such as transitivity, that the child does not yet have.

Again, to give a rule or demonstrate a principle by "show and tell" will not transmit the idea if the learner does not have the necessary logical structures to reconstruct or interpret what is "seen."

Memory is not a copy of reality.

Even at 8 years of age, or when children are in the third grade, they still have difficulty grasping the transitivity principle.[8]

---

[8]Jean Piaget and Barbel Inhelder, *Memory and Intelligence* (New York: Basic Books, Inc., 1973), p. 102.

ADDITION OF CLASSES

Purpose of Activity

The purpose of this activity is to determine when and how the child understands addition as an operation. This and the next activity will demonstrate that the learning of addition by many 6- and 7-year-olds is simply a verbal type of learning involving no true understanding of the number idea.

Addition will be considered here as applied to classes of objects—in this case, beads. The same idea is investigated in Activity 1-5, which considers the hierarchical relation of classes.

Materials Needed

1. Nine wooden beads—seven brown and two white (other colors may be substituted).
2. A piece of string or wire to use as a necklace.

Procedure

1. Ask the child what the beads are made of (wood).
2. Then, "What color is this [pointing at brown] and this [pointing at white]?"
3. "Are there more brown or more wooden beads?"
4. "If I made a necklace of wooden beads and a necklace of the brown beads, which would be longer? Why?"

Levels of Performance

Stage 1. This stage lasts on the average until 7 or 8 years of age. The child responds that there are more brown beads than wooden beads. Even if you ask, "Are the brown ones made of wood?" and "Are the white ones made of wood?" and repeat the question, "Then are there more brown beads or wooden beads?" the response is the same.

Psychologically speaking, the child can see or perceive either the whole set or its parts but not both simultaneously. Once the whole is considered in terms of its parts (brown and white), only the parts can be compared. The idea of the whole is lost. The child does not have **reversibility of thought,** allowing her to recombine the parts into the whole and consider the whole and its parts at the same time. This difficulty also exhibits the original basic problem of non-conservation of wholes—the child's thinking the part (brown) is larger than the whole.

The child, not having the necessary logical thought processes, answers on the basis of perceptual factors. The biggest set or class that she "sees" is brown, and she answers accordingly.

Stage 2. This is a transitional stage in which answers are at first like those in Stage 1, based on perception. But by trial and error the child arrives at the correct answer. If the interviewer asks such questions as, "Are the brown ones made of wood?" and "Are the white ones made of wood?" the child then may be able to correctly answer the question, "Are there more brown beads or wooden beads?" However, the child may very well be wrong again in a similar situation involving different objects but the same concept.

Five-year-old trying to decide if there are more flowers or more roses.

Stage 3. Answers are immediate and correct. These children have the necessary logical thought processes which are reversible:

Wooden = Brown + White

and also

Brown + White = Wooden

Children at this stage can then logically conclude that there are more wooden beads than brown beads. This is in contrast to Stage 1 children who have to answer on the basis of perceptual factors—they "see" more brown, so that is the answer.

Teaching Implications

This and the next activity are two of the most important in this book in that they expose a stage of unreadiness for addition in many children of 6 to 7 years of age. These children are learning addition without benefit of the necessary thought processes. Such learning is rote; it may cause negative feelings toward mathematics since what is being done is not understood.

ADDITION OF NUMBERS

Purpose of Activity

Joining a class of beads with another class of beads we still have a class of beads, but in joining a counting number with itself another number is produced. Psychologically then, is addition of numbers more difficult than addition of classes (the subject of the preceding activity)?

Psychologically, the ability to add numbers is found to develop at the same time as the ability to add classes, because the same logical processes of conservation of number and reversibility of thought are necessary in both activities. Does the child recognize the whole as being invariant in terms of different additive combinations such as 4 + 4, 7 + 1, 2 + 6?

Materials Needed

Dried beans.

Procedure

The child is shown beans arranged as follows:

$$\bigcirc \;\; \bigcirc \;\; \bigcirc \;\; \bigcirc \qquad\qquad \bigcirc \;\; \bigcirc \;\; \bigcirc \;\; \bigcirc$$

She is asked to imagine they are sweets. Today she is to have four in the morning and four in the afternoon. Tomorrow she can have the same number, but since she will not be as hungry, she can have one in the morning and seven in the afternoon. At this point, remove three of the beans from the set on the left and put them with the set on the right to represent the second day.

Then ask the child if she will have more on one day or the other or the same.

Levels of Performance

Stage 1. Until 6 or 7 years of age, the child thinks there is more on the second day because of the big lot (seven). Her answer is based on perceptual clues only, not having the necessary reversibil-

Dale, 7, chooses 4 + 4 rather than 1 + 7, saying the big lot (7) of candy would cause cavities.

ity of thought. Even if the experiment is repeated or she makes the transfer of the three beans herself, the answer is the same.

Stage 2.   Stage 2 is a transitional one in which the child first answers like the Stage 1 child but gradually comes to see or can be helped to see that the amount is the same. She does this by counting, although she shows surprise that the number is eight on both days.

Stage 3.   The answer is immediate and correct at around 7 years of age. This child is not concerned with the objects themselves as candy or beans but as *numerical units* which can be equated without regard for what the objects are (qualitative similarity). This is the transition from addition of classes to addition of numbers.

Teaching Implications

Having children repeat sums or memorize equations is learning only at a verbal level. Activities similar to the preceding one described reveal true understanding of a number, such as six, in terms of its addends—5 + 1, 2 + 4, 3 + 3.

The child must be able to combine parts to form a whole and realize equalities (4 + 4 = 1 + 7). She must also be able to divide a whole into parts and realize the equality. Symbolically, this activity involves the associative property:

$$4 + 4 =$$
$$(1 + 3) + 4 =$$
$$1 + (3 + 4) = 1 + 7$$

The necessary psychological conditions are (1) reversibility of thought and (2) consideration of three numbers simultaneously (the whole and its parts) and the quantitative relation that exists between them in regrouping addends (the associative property). (See also Activity 2-12.)

Subtraction as an operation is a part of this reversible system that Piaget describes as addition. In fact, the greater difficulty children have with subtracting is due to the necessity for reversible thought. As addition is often taught as a one-way or to-the-right operation, 2 + 3 = 5, and not also as 5 = 2 + 3, this difficulty is not immediately apparent. But in a missing-addend subtraction, such as 3 + □ = 5, the child must look to the right and then back to fill in the missing addend. Many children are unable to do this, saying the answer to 3 + □ = 5 is the number 8.

Psychologically, addition and subtraction are one reversible system. (See also Activity 2-10.)

ADDITION AND SUBTRACTION (MAKING QUANTITIES EQUAL)

Purpose of Activity

The purpose here is to study children's ability to make use of addition and subtraction to solve problems involving unequal quantities.

Materials Needed

Two sets of counters, one of 8 and one of 14.

Procedure

Show the child the two sets and ask her to make them equal.

Levels of Performance

Stage 1. At this stage, from 4½ to 6 years of age, the child does not understand that addition and subtraction compensate each other as a reversible system. As she adds counters to the smaller set, she is not aware that she has reduced the larger. She may not even use the logic of number and simply make a global comparison— "They are the same because they 'look' the same size" (occupy about the same amount of space).

Stage 2. The child at this stage has an intuitive grasp that in adding to one set she is subtracting from the other, but equalizing the sets is still a trial-and-error process. As she moves the objects from one set to the other, she judges by physical appearance. She may then count to make sure.

Stage 3. Children aged 5½ to 7 have an immediate grasp of the necessary addition and subtraction. They realize that 8 and 14 can be equalized by adding 3 of the 14 to the 8, that the 8 becomes 11, and that 14 − 3 also becomes 11. They solve the problem at the numerical level without being tied to the physical arrangement of the counters as are the Stage 2 children.

For example, both a Stage 2 and a Stage 3 child begin by arranging the set of 8 as follows:

```
•   •
•   •
•   •
•   •
```

They then begin arranging the set of 14 in the same pattern beside the set of 8.

```
•   •   •   •
•   •   •   •
•   •   •   •
•   •   •   •
```

The 6 left over are divided by placing 2 more in each set and then 1 more in each set.

The difference between the Stage 2 and the Stage 3 child is that the former is tied to the physical layout, making judgments *after* the division is made and relying on how the figures look after the division. If the shape of one of the sets is changed, she is not sure of the equality. Hers is not a numerical division, but a physical one.

The Stage 3 child, on the other hand, knows from the beginning that if one set is larger than the other, she must therefore divide the difference (6). This she does at the numerical level and then demonstrates by arranging the counters.[9]

Teaching Implications

This activity identifies children who have an operational understanding of addition and subtraction as contrasted to those attempting to solve such problems without the necessary mental operations.

As summed up by Piaget, "numerical addition and subtraction become operations only when they can be composed in the reversible construction which is the additive 'group' of integers, apart from which there can be nothing but unstable intuition."[10]

---

[9]Jean Piaget, *The Child's Conception of Number* (New York: W. W. Norton & Company, Inc., 1965), p. 194.

[10]Ibid., p. 195.

THE COMMUTATIVE PROPERTY
OF ADDITION

Purpose of Activity

What is children's understanding of the commutative property of addition—that the order of adding numbers does not change the sum? For example, when do children realize that $2 + 7 = 7 + 2$?

Materials Needed

Two sets of objects, such as 2 yellow toy cars and 7 blue cars.

Procedure

Ask the child to line the cars up in a row with all the cars of one color followed by all the cars of the other color. Put a parking marker at both ends of the row. Then remove the cars from the row. If the

Checking to see if a row of two "Mickey Mouse" dolls followed by three dwarfs is the same length if the order is reversed.

73

child began his row with yellow cars, call that to her attention and ask her to do it again beginning with the blue cars. After she has placed two or three cars, stop her and ask, "Do you think this row will reach to the same parking marker, or not as far, or farther? Why?"

A variety of other materials could be used for the same purpose—dolls of two types, for example.

Also, to investigate the idea further, use continuous sets, such as liquid to which food coloring has been added.

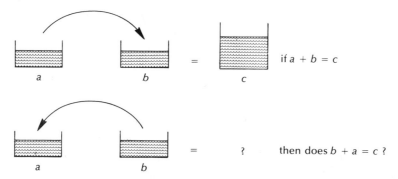

Levels of Performance

Children will be successful with these tasks on the average at 7 to 8 years of age. The tasks involve the concept of reversibility which develops around 7 years of age.

Teaching Implications

Present practice of teaching addition in the first grade ignores the necessity of reversible thought as a basis for understanding the operation. It is one thing to teach that 3 + 2 is equal to 5 and that 2 + 3 is equal to 5 and have children memorize that the response is 5 in both cases; but the activity just presented demonstrates that teaching the concept is not the same as teaching a response.

When children *do* have reversible thought, the commutative property simplifies teaching the 100 basic addition facts, because for every addition fact such as 7 + 2, they also know 2 + 7.

THE ASSOCIATIVE PROPERTY OF ADDITION

Unlike seriation and transitivity, associativity involves structures that cannot easily be fixed in the memory.[11]

Purpose of Activity

In this activity children's understanding of, and memory of, the associative property of addition will be explored.

Materials Needed

1. Two sets of counters—one of 8 and one of 14.
2. Four glasses and some liquid, as illustrated on the following page.

Procedure

In Activity 2-10 on addition and subtraction, the child was asked to equalize sets of 8 and 14 counters. One way of doing this could involve thinking of the 14 renamed as $8 + 6$ and dividing the 6 left over between the two sets. Expressed numerically,

$$8 + 14 =$$
$$8 + (6 + 8) =$$
$$8 + (3 + 3 + 8) =$$
$$(8 + 3) + (3 + 8) =$$
$$11 + 11 =$$

The whole amount in the two sets $8 + 14$ is not changed by regrouping the set of 14 from $6 + 8$ to $3 + 3 + 8$ and associating one of the addends, 3, with the addend, 8.

---

[11]Jean Piaget and Barbel Inhelder, *Memory and Intelligence* (New York: Basic Books, Inc., 1973), p. 117.

Does it make any difference, in adding any three or more numbers, which pairs of numbers you group or add first? For example,
does

$$(3 + 2) + 5 = 3 + (2 + 5)?$$
$$5 + 5 = 3 + 7$$
$$10 = 10$$

The parentheses indicate which number pairs are associated first,
and it is seen that the different associations produce the same sum.
Since this is true for any whole numbers a, b, and c, the idea is called
the **associative property** and is usually stated as

$$(a + b) + c = a + (b + c)$$

To explore this idea at the physical or concrete level, Piaget used a
process of pouring liquids.

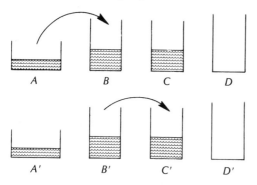

The child is shown the display as illustrated, and is told that each
glass holds the same amount as the one above (or below) it. The
interviewer then pours the liquids as indicated by the arrows.

Will you get the same amount if the liquids are grouped differently; that is, do you get the same amount by pouring (indicate by
pointing, not naming letters) A into B and the result into C as by
pouring A' into B' and C' combined? The child is then asked if these
(B and C) are poured here (pointing at D), and if these (pointing at A'
and C') are poured in here (pointing at D'), will you get the same or
more in one than in the other? If the child thinks there will be more
in one, ask which one and why. Symbolically, this can be represented as

$$(A + B) + C = A' + (B' + C')$$

The interviewer then pours the liquid into D and D', so that the child can see that they do contain the same amount, and the child is asked if she can explain why.

If the interviewer is interested in how this activity relates to memory, the child is asked one week later if she can remember what she did and if so, to draw it. The materials are then brought out and the child asked to do what was done the week before. Finally, she is asked if it is the same here and here (meaning D and D'). The same procedure is repeated six months later to see if, in the interim, the child has acquired better cognitive processes to resolve the problem. (For a full discussion see Piaget and Inhelder's *Memory and Intelligence,* Chapter 6.)

Levels of Performance

Piaget finds five levels of performance as far as memory is concerned. Because the description of these levels is so lengthy, it will not be attempted here. (See *Memory and Intelligence,* pages 118–23.)

Children's understanding of conservation of quantity seems to be a prerequisite to understanding associativity. The child must understand that if the liquids are the same in A as in A', in B as in B', and in C as in C', then pouring them into different-shaped containers does not change the amount.

At 7 to 8 years of age, children do have a stable concept of conservation of quantity; thus, they are also able to remember some of the associations made (how the liquids were poured).

At 8 to 9, some of the children are able to use the idea of the associative property (that the different associations make no difference and that, therefore, there will be the same amount in D as in D').

But Piaget points out that in the case of associativity, subjects 6 to 9 years lack adequate schemata, having only conservation schemata. Differentiated associativity does not emerge until later, and is not fully consolidated until 11 to 12 years of age.[12]

Teaching Implications

The associative property is more difficult for children than many elementary math textbook writers realize. Some first-grade books include this idea. It would seem more appropriate to delay intro-

---

[12]Ibid., p. 126.

ducing the idea until the child reaches the second or third grade, realizing as described in the preceding paragraph that a full under- standing of associativity does not occur until 11 to 12 years of age.

A further study of associativity can be made by means of Activity 2-10 or Piaget's *The Child's Conception of Number,* page 190.

While the apparatus for this activity is set up, it may also be in- teresting to investigate the commutative property of addition; that is, does the order of adding liquids affect the amount?

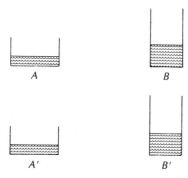

If A and A' contain the same and B and B' contain the same, do we get the same amount pouring A into B as B' into A'? (In conducting the activity, do not use letters but point at the appropriate con- tainer.)

MULTIPLICATION OF NUMBERS

Purpose of Activity

The purpose of this activity is to determine when children are able to solve problems involving multiplication of numbers.

The ability of children to make a matching or one-to-one correspondence between sets was studied in Activity 2-6 on conservation of number. From the psychological standpoint, a one-to-one correspondence is an implicit multiplication, for if two sets are in one-to-one correspondence, then the total number is twice or two times the number in each set.[13]

In this activity more than two sets will be considered in order to study children's ability to predict the total number as two, three, or four times the number in each set.

Materials Needed

Cardboard cutouts of 10 blue flowers, 10 yellow flowers, 10 vases.

Procedure

Ask the child to put a blue flower in each vase. Then remove the flowers and ask her to put a yellow flower in each vase. After these are removed, ask her if there are the same number of blue flowers as yellow flowers. And then, "If we put all the flowers in the vases with the same number in each vase, how many would there be in each vase?"

---

[13]Jean Piaget, *The Child's Conception of Number* (New York: W. W. Norton & Company, Inc., 1965), p. 204.

Levels of Performance

Stages 1 and 2.   Between 5 and 6 years of age, the child can make a one-to-one correspondence of flowers to vases, but if one set of flowers is bunched up afterwards and the other set spread out, she thinks there are more flowers in the spread out set. She does not have the conservation of number concept.

Also, the child does not have the transitive idea:

If the number of

(blue flowers) = (vases)

and

(yellow flowers) = (vases)

then

(blue flowers) = (yellow flowers)

This **transitive** property of the equals relation may be expressed symbolically as

if $a = b$ and $b = c$, then $a = c$

The Stage 1 child cannot apply this logical idea.

At Stage 2, the child gradually succeeds by repeating the experiment or through the interviewer's questions such as "How many blue flowers were in the vases?" and "How many yellow flowers were in the vases?" But perceptual factors still influence her—as one set is spread out more, for example.

Stage 3.   At 6 to 7 years of age, the child can immediately use the logic of the transitive property and conclude that since one blue flower went in each vase and one yellow flower went in each of the same vases, there must be the same number of blue flowers as yellow flowers.

The Stage 3 child can also answer the second question successfully: "If we put all the flowers in the vases with the same number in each vase, how many flowers are in each vase?"

If a third set of ten flowers is introduced and the question repeated, she can predict that each vase will hold three flowers.

If a set of straws is introduced with the restriction that only one flower will go in each straw, and the child is asked, "How many straws will we need for all the flowers?" she can predict how many. The Stage 2 child would have to go through the process of pairing straws and flowers to find the answer and then realize why the answer is what it is.

Teaching Implications

Multiplicative operations are seen to develop at the same time as additive operations in the mind of the child. Yet multiplication is not studied in comprehensive fashion until the third or fourth grades in most schools. This is probably because there are 100 addition facts to learn, and although they may not be easier, it takes time to learn them. Part of the problem may be beginning addition too soon and multiplication too late.

BEGINNING DIVISION

Purpose of Activity

The purpose of this activity is to study children's operational concept of equal division.

Materials Needed

A set of 18 counters.

Procedure

The child is asked to divide the counters "so that you and I have the same amount."

Levels of Performance

Stage 1. A child at this stage arbitrarily divides the counters by grabbing two handfuls and separating them into two piles. Asked if she and the interviewer each have the same, she judges by the space occupied by each pile.

Another Stage 1 child begins on a numerical basis, using a one-for-me–one-for-you approach. She divides the counters correctly in this manner, but when asked if the amounts are equal, she says there is more in one pile which happens to be spread out more. She then tries to make them the same by compressing the more spread out set.

"But are they the same?"

"No."

Stage 2. The child aged 5 to 6 divides the counters in pairs of "one for you and one for me" and then arranges the two sets into some geometrical pattern to determine if they are equal. For example,

```
•  •  •      •  •  •

•  •  •      •  •  •

•  •  •      •  •  •
```

Stage 3. This child, 6 to 7, divides the counters in the same manner as the Stage 2 child, but knows the results will be equal based on the one-to-one correspondence in the division. She does not have to judge on what the results look like; that is, in terms of the same geometric designs or the same spaces occupied.

Teaching Implications

Readiness for the operation of division appears at about the same time as for its inverse operation, multiplication, and readiness for multiplication appears at about the same time as that for addition.

As described by Piaget, "Additive and multiplicative compositions [which include division] . . . are . . . correlative, and the mastery of the one implies that of the other."[14] Thus more emphasis might be placed on integrating multiplication and division activities with addition and subtraction, rather than delaying multiplication and division for one or more years.

---

[14]Jean Piaget, *The Child's Conception of Number* (New York: W. W. Norton & Company, Inc., 1965), p. 198.

FRACTIONS

Purpose of Activity

The purpose is to explore children's understanding of the quantitative relations involved in dividing a whole into equal parts.

Materials Needed

1. Paper cut into various shapes, such as circles, squares, and rectangles.
2. Six dolls.

Procedure

Show the child a circular piece of paper and two dolls and ask, "If this were a cake, could you cut it so that each doll would get the same amount? Can you show me with this pencil how the cake should be cut."

Vary the procedure with other shapes, such as a square and a rectangle.

Also use the same procedure with three dolls, then six dolls.

Levels of Performance

Stage 1. In children under 4 years of age, there is no plan and the divisions are not equal. The child is unable to relate the part to the whole once the cake is cut. The parts are now less or more than the original cake.

Stage 2. At Stage 2A, 4 to 6 years, children can divide small cakes into two equal parts, but not into three equal parts.

At Stages 2B and 3A, 6 to 7 years of age, children can divide into three equal parts, but it is a trial-and-error process; for example, drawing a line and then erasing it to redraw the line in a different position. These children do understand conservation of the whole—that once the cake has been cut it is still the same amount of cake.

Stage 3. At Stage 3B, around 10 years of age, children can with

Dividing a "cake" for three dolls.

some assurance tackle the problem of dividing the cake into six equal parts. They may, for example, first divide it into three equal parts and then divide each of the three into two equal parts.

Teaching Implications

Children are often introduced to fractions by being "shown" something divided and "told" the name of each part. Then they are given pictures divided into parts and asked to name the parts.

Children should do the dividing as well as naming the resulting fractional part. As this activity demonstrates, it is not as easy as one might think.

RATIO AND PROPORTION[15]

Purpose of Activity

The purpose of this activity is to determine children's ability to understand the idea of a **proportion,** for example, that 1/2 = 2/4. The model used is speed as a relation of distance to time.

Materials Needed

1. Two toy cars.
2. A stopwatch.
3. A meter stick.
4. A pencil.

Procedure

The idea of speed or how fast something is going can be expressed as a **ratio** of distance to time (2 meters in 4 seconds or 2:4).

To compare the speed of two moving objects using the distance each traveled divided by the time each traveled involves the idea of proportion. For example, if one car goes 2 meters in 4 seconds and another 1 meter in 2 seconds, the speeds are the same—2:4 or 1:2.

Equal ratios such as this can be expressed as a proportion

$$\frac{2}{4} = \frac{1}{2}$$

Can children use this idea to determine whether or not the speeds are the same?

Using a clock with a second hand and two toy cars, give the child the clock and ask her to tell you when to start and stop a toy car so that it travels for 2 seconds. Then draw the path the car traveled as a line segment 1 unit in length (1 meter or decimeter, for example). Record the time of 2 seconds.

---

[15]See also Chapter 7 on chance and probability.

Marge, 9, can successfully compare the speed of two cars based on ratios of distance to time traveled.

Repeat the procedure with another toy car traveling on a path parallel to the first car, but this time ask the child to tell it to stop after 4 seconds. Make the car travel 2 units in length during this time period.

Now ask the child, "Did one car go faster or did they both travel at the same speed?"

As another activity, move a car along a path such as 4 centimeters in 4 seconds. Then on a parallel path move a second car the *same* distance in 5 seconds. Mark the paths by drawing two line segments and indicate distance traveled on each. Have the child use a stopwatch to determine the time traveled for each car. Record the time traveled by each path.

$$\overline{ \overset{\text{4 cm}}{} } \quad \text{4 sec}$$

$$\overline{ \overset{\text{4 cm}}{} } \quad \text{5 sec}$$

Ask the child if one car went faster and why.

As a third activity, make the distance traveled different, such as

4 centimeters and 5 centimeters, but the time traveled the same. Again ask if one went faster and why.

Still another activity would be to make both distance and time different, using simple ratios such as 4 centimeters in 2 seconds and 2 centimeters in 1 second. Again ask if one car went faster and why.

4 cm
—————————————————————   2 sec

2 cm
——————————   1 sec

Levels of Performance

Stage 3A (Ages 7–9). For a car traveling 4 centimeters in 4 seconds followed by another car traveling 4 centimeters in 5 seconds:

"Did the cars take the same time?"

"Yes."

"Did they cover the same ground?"

"No, the second one went farther." (Thus confusing distance with time.)

Another child is questioned (same stage):

"Same distance?"

"Yes."

"Same time?"

"No, 4 seconds and 5 seconds."

"Does one of them go faster?"

"The one in 5 seconds."

"Why is it faster?"

"Because it took more seconds." (Thus concluding the one that took longer went faster, even though the distance was the same.)

For the second activity in which distances are different, such as 4 centimeters and 5 centimeters in 4 seconds, an 8-year-old is questioned:

"Do they go the same distance?"

"No."

"Do they take the same time?"

"Yes."

"Does one go faster?"

"Yes, the one which gets farther ahead."

"Why was it faster?"

"Maybe that's the reason why."

This child is then shown a car going 2 centimeters in 1 second and another going 4 centimeters in 2 seconds:

"Did one go faster?"

"The one taking 2 seconds because it went a longer distance."

"Are you sure?"

"No, it's the other one because it went only a little way and takes less time."

This 8-year-old has no idea of these two speeds as ratios of 4/2 and 2/1 being the same.

Stage 3B (Ages 10–11). These children at first make mistakes characteristic of Stage 3A, but they are able to correct their errors with further questioning. While they can compare speeds if the times or distances are the same, they are still unsuccessful if both distances and times are different. For example, they have difficulty when asked, "If one car travels 5 centimeters in 6 seconds and another 4 centimeters in 3 seconds, is one faster?" They may solve such problems with simple ratios like 2 to 1, but are unable to generalize the idea.

Stage 4 (Ages 10–12). In this Formal Operations stage, measurement problems are solved systematically, even when times and distances are different. The idea of a proportion is used even though the children may not know how to make the calculation correctly.

For example, for one car going 2 centimeters in 4 seconds and another 6 centimeters in 8 seconds:

"The second one is faster because it is not a half." (Meaning 2 centimeters is half of 4 seconds but 6 centimeters is more than half of 8 seconds.)

Comparing a distance of 16 with a time of 6 and a distance of 15 with a time of 5, another youngster responds, "You would divide 16 by 6 and 15 by 5."[16]

Teaching Implications

The idea of a proportion as

$$\frac{a}{b} = \frac{c}{d}$$

---

[16]Jean Piaget, *The Child's Conception of Movement and Speed* (New York: Ballantine Books, Inc., 1971), p. 242.

may be expressed in terms of speed as

$$\frac{distance_1}{time_1} = \frac{distance_2}{time_2}$$

Such a proportion is very useful in solving problems of this type. For example, If a car goes 90 miles in 2 hours, how far will it go in 5 hours?

$$\frac{90}{2} = \frac{x}{5}$$

Proportions are not usually introduced into the curriculum until the fifth or sixth grades. This corresponds fairly well with the necessary developmental level (formal operations) of the children. However, 10 to 12 years of age is an *average* for the beginning of the formal operational level; therefore, the student who is below average may not be ready.

Possibly more important is the implication of readiness for understanding addition of unlike fractions. Unlike fractions are taught using the idea of equivalence classes; for example, to solve a problem such as 1/2 + 1/3, the equivalence classes for 1/2 and 1/3 can be displayed as

$$\frac{1}{2}, \frac{2}{4}, \frac{3}{6}, \frac{4}{8}, \cdots$$

$$\frac{1}{3}, \frac{2}{6}, \frac{3}{9}, \frac{4}{12}, \cdots$$

Does the child realize that 1/2 and 3/6 or 1/3 and 2/6 do in fact name the same number? The equality is often not realized in terms of speed, as the activity we have just described demonstrates. For example, a speed of 1 centimeter in 2 seconds is not considered to be the same speed as 2 centimeters in 4 seconds.

Piaget described his findings:

> We have found that the classical notion of speed as a relationship between the spatial interval [distance] and the temporal duration [time] appears very late in child development, about 9 or 10 years of age.[17]

The necessary intellectual constructions to use the idea of a pro-

---

[17]Jean Piaget, *Genetic Epistemology* (New York: Columbia University Press, 1970), p. 62.

portion in mathematics are not present until the formal operational level—usually at around 11 or 12 years of age. While only speed has been considered here, children often have the same developmental problems in studying probability, comparing similar figures in geometry, map making, and studying percentage, since all of these activities involve understanding proportions.

3

Space Orientation

Children's concepts of space are not the same as those of the adult. Consider the little girl who goes to a friend's house with her father. Knocking on the door she says, "Get behind me, Daddy, so they won't see you."

The exploration of space may be thought of as a geometrical activity. Children explore space from the time they are born, long before they explore number. This exploration can serve as an excellent basis for abstracting number ideas. According to Piaget, "Motor activity [is] of enormous importance for the understanding of spatial thinking."[1]

The following activities are included in this chapter on space:

3-1.	Drawing (Copying) Geometrical Figures
3-2.	Perceptual and Representational Space
3-3.	A Horizontal and Vertical Reference System
3-4.	Locating Objects in Space
3-5.	Making a Map
3-6.	Perspective
3-7.	Subdividing a Line
3-8.	Reciprocal Implication and Equal Angles

---

[1]Jean Piaget and Barbel Inhelder, *The Child's Conception of Space* (New York: W. W. Norton & Company, Inc., 1967), p. 13.

DRAWING (COPYING) GEOMETRICAL FIGURES

Purpose of Activity

This activity is to explore children's ability to copy basic geometrical figures.

Materials Needed

1. A card with figures as shown on the next page.[2]
2. A pencil.

Procedure

Ask the child to draw a man from memory both to put him at his ease and to get some idea of his drawing ability. Then ask him to make a copy of each of the figures on the card. These figures make distinctions between **topological** and **Euclidean** concepts of space.

Levels of Performance

Stage 0. At 2½ to 3 years of age, drawings are pure scribbles.

Stage 1A. At 3 to 4 years of age, the scribbles vary according to the model being copied. Open shapes such as Models 20 and 21 can be distinguished from closed shapes such as Models 4, 5, and 6.

Stage 1B. The drawings of these children, also 3 to 4 years of age, Piaget refers to as the first real drawings bearing some resemblance to the figures being copied. Definite distinctions are made between open and closed figures, but the closed figures are not distinguishable from each other. A circle, triangle, and square may all be drawn as an irregular closed figure.

Stage 2. Children aged 4 to 6 show progressive differentiation between the basic shapes, such as circles, squares, and triangles, in their drawings. Curved shapes are distinguished from straight-sided

---

[2]Jean Piaget and Barbel Inhelder, *The Child's Conception of Space* (New York: W. W. Norton & Company, Inc., 1967), p. 54.

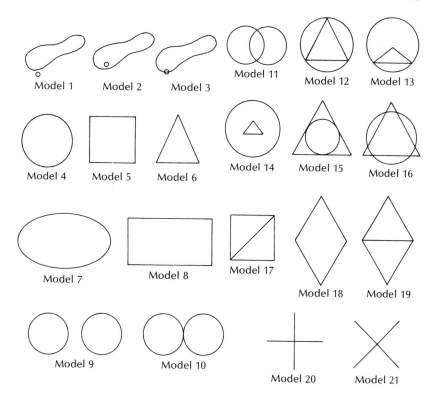

Model 1   Model 2   Model 3   Model 11   Model 12   Model 13

Model 4   Model 5   Model 6   Model 14   Model 15   Model 16

Model 7   Model 8   Model 17   Model 18   Model 19

Model 9   Model 10   Model 20   Model 21

ones, but little distinction is made at first between the curves or between the straight-sided figures.

A square and rectangle may appear to be the same. Gradually, differentiation is made in terms of number of angles. A triangle, for example, may be different from a square; a circle, different from an ellipse. The two crosses, Models 20 and 21, are drawn differently, indicating the recognition of difference between vertical, horizontal, and oblique. Circumscribed figures such as Models 13, 14, and 15 are drawn correctly as to shape, but the points of contact are not properly represented; that is, as touching, not touching or overlapping.

Stage 2B. The rhombus, Model 18, is now drawn correctly. Circumscribed figures such as Models 12, 13, and 15 (but not 16) are gradually mastered.

Stage 3. Children at this stage, 6 to 7 years of age, are able to draw all of the models correctly.

Teaching Implications

This activity provides a basis for determining in a clearly defined way what progress children are making in their ability to realize, as well as copy, basic spatial distinctions.

Children first make topological distinctions between figures. They distinguish "open" figures (Models 20 and 21) from "closed" figures, such as Models 4, 5, and 6. They are also able to make the topological distinctions in Models 1, 2, and 3: They can copy the small circle "outside," "inside," and "on the other figure." Some children, not functioning at the topological level, cannot even do this. In drawing a human figure, they may draw an eye outside the head or legs not attached to the body.

Later, children make Euclidean distinctions which show correctly the sides or angles; the length of sides; the size of angles; and vertical, horizontal, and oblique concepts. Thus they differentiate between such Euclidean shapes as squares, parallelograms, and trapezoids.

Most children will be ready at first-grade level to learn the basic shapes.

PERCEPTUAL AND REPRESENTATIONAL SPACE

Purpose of Activity

Knowledge of space or the world around us develops in two ways—first, through what we perceive or what our senses tell us. Piaget refers to this concept as **perceptual space.** But our minds do not operate like cameras—photographing what our eyes see. Instead the mind must reconstruct or **represent** mentally such sensory data as obtained by touching or seeing. What the mind reconstructs is referred to as **representational space,** the second concept.

The purpose of this activity is to study children's ability to convert sensory or perceptual impressions of an object into mental representations of that object which, of course, is necessary for full understanding. What the mind "represents" may be and often is different from what is "seen" or "felt" by small children.

Materials Needed

1. A set of familiar objects such as a ball, comb, spoon, pencil.

2. A set of cardboard topological forms of a size convenient to handle.

3. A set of cardboard Euclidean shapes approximately 3 by 4 inches in size.

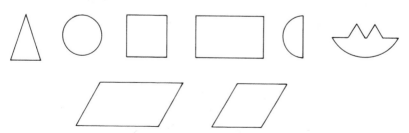

4. A cardboard screen with two holes in it for the child to put his hand through while handling objects he cannot see.

97

5.   On a poster, pictures of objects the child is to handle (1–3).

Procedure

Ask the child to put his hands through the screen. Then, one at a time, give him objects to handle, asking him to point at the object on the poster.

For children 2 to 3 years old, use the familiar objects. For children 4 or older, begin with the topological figures. Piaget finds children able to make topological distinctions first—whether a figure is closed or open, has one or two holes, is intertwined or simply overlapping. It is more difficult to make Euclidean distinctions, such as noting the differences among triangles, squares, rectangles, parallelograms, and rhombi.

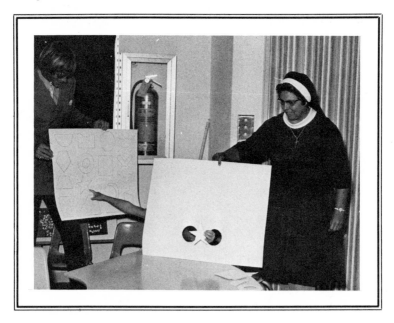

Levels of Performance

Stage 1.   Familiar objects such as the comb are recognized, but children under 4 years of age are not coordinated in their action of handling the object. Coordination, of course, is necessary to construct a mental representation of the object.

Exploration is passive. No distinction is made between such Euclidean figures as the circle and square (since both are closed figures). Toward the end of Stage 1, distinctions are made between the topological figures such as whether the figure is open ℭ or closed ◎ .

Stage 2. Usually, children from 4 to 6 years of age can distinguish between curved Euclidean shapes and those with straight sides, but no distinctions are made between such rectilinear figures as squares and rectangles. Neither are the curvilinear figures, such as circles and ellipses, differentiated from each other.

Explorations of the figures is more methodical, however; such shapes as the triangle and square begin to be differentiated by number of angles.

Stage 3. Beginning at approximately age 6, children are operational as far as basic Euclidean shapes and even more complex figures such as a cross are concerned. Account is now taken of order and distance between points on the figure.

Exploration is methodical. Furthermore, although it begins with perceptual activities as in the other stages, this stage now uses an operational method of *grouping* the elements perceived, such as corners or angles and the distances between them. This grouping is dependent on beginning at a fixed point on the figure, tracing the outline and being able to reverse the process, tracing back to the starting point, mentally grouping those sensory impressions into a coordinated whole at the representational level. This, children in the first stages cannot do.

Teaching Implications

Teachers need to know the stages through which children go in developing the ability to consider geometric ideas. Many teachers may also not be aware of the fact that understanding space is not just a perceptual activity. For example, teaching what a triangle is by the "show-and-tell" method does not at all ensure that the child can mentally represent the idea of a triangle and hence understand it.

Perceptual activities are the starting point but are not sufficient. Mental coordinating actions of these sensory impressions must take place. These actions must often be based on physical or motor action, such as tracing and retracing a figure with one's fingers. For some children the necessary mental coordination may not occur until 9 or 10 years of age.

Also, the task of picking a figure from a collection that the child has handled but not seen is less difficult than having to represent or draw the figure. However, children should also have activities in representing figures by drawing them when they are able. If motor coordination is poor, the child may be asked to make the figures out of matchsticks.

A HORIZONTAL AND VERTICAL REFERENCE

Purpose of Activity

The two preceding activities involved perspective or how an object "looks" from different points of view. In Euclidean geometry, we are concerned with the relations of one object to another in space; for example, as in making a map. Relative to each other, are the objects up, down, right, left, parallel, and how far? A familiar problem of this sort might be finding one's car in a parking lot or trying to follow directions in a strange city.

The person "good at directions" is so because he or she has a stable **reference system,** such as imaginary vertical and horizontal lines or axes. Each object is placed in relation to these axes, so that the objects can be related to each other. Such a person uses natural objects, such as a road or the sun or the north star, as a point of beginning for one axis and imagines the other axis at right angles to this.

A child needs such a reference system in order to make a map of his neighborhood or to have a mental overall coordinated view of his surroundings. What meaning does horizontal and vertical have for the child as a basis for establishing a reference system?

Materials Needed

1. Two wide jars with straight sides, one half filled with colored liquid.
2. Paper cutouts of jar, tilted at a 45-degree angle from the vertical.
3. A crayon.

Procedure

Show the child the jar with colored liquid in it. Tilt the other jar 45 degrees to the right, and ask the child to show with his finger where

the top of the water would be if that jar had water in it. Then tilt the
jar to the left and repeat the question.

The procedure can be varied by using sketches of bottles tilted 45
degrees and asking the child to color inside where he thinks the
water would be. The bottle containing the liquid is then tilted and
the child asked to check his drawing against what actually happened.

The idea of vertical can be presented by drawing a hill on a piece
of paper and asking the child to draw a house, pole, or tree on each
side of the hill.

Levels of Performance

Stage 1.   Up to 4 or 5 years of age, the child scribbles to represent
water in the jar.

Stage 2.   This stage, from ages 5 to 9 approximately, lasts longer
than for many of the other activities. At first the child splits the water
as he draws, drawing half of it parallel to the base of the jar and half
parallel to the side, or he places it all parallel to the base or side.

He is using the idea of parallel but does not think of the jar in
relation to the objects around him, such as the horizontal table or
floor. After the child sees the jar containing the liquid tilted, he may
be able to correct his error for that particular tilt. But the next day, or
for another angle of tilt, he is no longer able to solve the problem.
He has no stable reference system.

Stage 3.   Around 9 years of age, the child usually develops the
necessary horizontal and vertical reference systems which allow him
to draw correctly and immediately the position of the water, regardless of the angle of tilt.

Teaching Implications

This very simple experiment reveals the fundamental starting point
for a child's being able to correctly locate himself and other objects
relative to each other. Does he have a reference system? If not, what
meaning can a map of his neighborhood have, for example, unless
he can constantly relate it to a stable reference system as he makes
his way home or to a friend's house.

Piaget found that young children have no such system. When they
get to a landmark, such as the candy store, they may be able to point
to the next point on the way home. However, this is a sensorimotor

or stimulus-response schema. The children have no overall coordinated idea of the various landmarks on the way home.

The implications of this finding for readiness in studying geography are obvious. To make maps the mathematical idea of ratio and proportion will also be necessary in order to reduce to scale. Piaget found this ability does not occur in children until around 11 or 12 years of age, the formal operational level. Fortunately, that is the age group currently being taught the idea of ratio and proportion.

Child's idea of vertical *as represented by a chimney. Topologically, windows are correctly placed "inside" the house, but incorrectly in terms of Euclidean position.*

LOCATING OBJECTS IN SPACE

Purpose of Activity

The purpose of this activity is to study children's ability to locate objects in space relative to each other.

Materials Needed

1. Two identical models of open country with some basic landmarks; for example, a stream or road, a mountain, a house, and a barn.
2. Two small dolls (to the scale of the models).
3. A screen to block the view from one model to the other.

For older children, pencil-and-paper layouts may be used.

Misty, 7, has difficulty copying a model layout.

Procedure

First show the models without the screen with a doll on one model. Ask the child to place his doll in the same spot on the other model.

Then move the doll to a new position and ask the child to do the same. After several movements, rotate the child's model 180 degrees and repeat the procedure.

Now place a screen between the models. Allow the child to look at either model, but do not let him see both at the same time. It may be helpful to ask the child to close his eyes and describe what he is trying to do.

Levels of Performance

Stage 1.   Although children from 2 to 4 years of age cannot perform this task, they do have a good sense of direction and can point out from the window important reference points. While walking home, they can show the way from one reference point to the next; however, the entire route is not a coordinated system for them, but rather, a sensorimotor-type knowledge. ("When I 'see' the church I turn this way.") If you make an about-face with them, they become confused. Their "way home" is a rigid type of knowledge.

To locate the doll, children at Stage 1 use the topological relations of proximity and enclosure. The doll may be correctly placed "near" the house or "in" the yard, but the child is unaware of the Euclidean relations of right, left, in front or in back of the house.

Stage 2.   From 4 to 7 years of age, there is a progressive ability to locate the doll correctly. Now the doll may be correctly placed "behind" the house, but to the left rather than to the right of the barn. More than one relationship may be correctly established, thus performing a "logical multiplication." Knowledge is still of a perceptual or motor sort. Space is still not a coordinated whole.

While younger children often can do mirror writing or perceive inverted pictures better than older children, the constant reversal of left-right and before-behind relationships in this activity confuses them.

Stage 3.   At Stage 3, 7 to 8 years of age, the child can coordinate relationships and correctly place the doll in a number of positions, even when one model is rotated 180 degrees relative to the other. The law of associativity allows them to reach the same point by a variety of routes.

Teaching Implications

This simple activity provides a keen insight into the child's ability to consider positions in space relative to each other. This is, of course, important in such areas of study as geography. Space for the child does not become a coordinated and objective whole until around 9 years of age.

Part of this problem was examined in the preceding activity. In order to represent Euclidean spatial ideas, an abstract coordinate or reference system must be formed in which the objects can be positioned. This also occurs at approximately 9 years of age.

The next activity goes a step further to investigate children's ability to make a map. This involves reducing to scale, or the use of proportion. (See also Activity 2-1 for linear relationships between objects.)

MAKING A MAP

Purpose of Activity

The purpose of this activity is to determine children's ability to make a map.

Materials Needed

1. A model with some objects on it, such as a house, barn, car, tree, and road.
2. A sheet of paper smaller than the model.
3. A pencil.

Younger children may be given a duplicate set of smaller objects and a smaller piece of cardboard on which to place them, rather than having them draw a map with paper and pencil.

Procedure

Show the child the model. Then give him the sheet of paper, which is smaller than the model, and ask him to draw a map of the model.

After the child has drawn his map, remove the objects from the model and ask the child to place them back where they were, using his map as a guide.

As a variation, the child may be asked to draw the model from different vantage points, such as the barn.

Levels of Performance

Stage 1. Children under 4 years of age use the topological relation of proximity, placing the objects near where they should be but not in correct relation to other objects.

Stage 2. At 4 to 6 or 7 years of age, the child is still unable to coordinate or multiply relationships of distance and order. He may correctly coordinate the objects in a small group with each other, usually in pairs, but he does not correctly relate groups to other groups. Distances are ignored. He also may use only a part of his map or page to locate all the objects.

Stage 3. In the child from 7 to 10 years of age, a system of reference is gradually built through logical multiplication (to the right of the house, behind the tree, in front of the barn). Left-right and before-behind relations are observed. During the first part of Stage 3, distances are still only partially taken into account. Proportions are ignored in terms of the size of the objects drawn, and objects drawn on the map are the same size as those on the model. Toward the end of Stage 3, the child masters the distance and proportion problems involved. He draws the objects smaller on the map and locates them correctly in their relation to each other, both as to direction and to distance, which involves a proportion judgment.

Stage 4. All that remains to be accomplished at this stage is to use a purely schematic plan—"substituting for the drawing of material objects a diagram of the area on which their positions are established by exact measurement."[3] This is accomplished at Stage 4, formal operational, around 11 to 12 years of age.

Exact measurement will require the ability to use a proportion, selecting a scale or ratio such as 1 centimeter on the map for 5 centimeters on the model. If, then, the distance between the house and barn on the model is 12 centimeters, the problem can be solved as

$$\frac{1}{5} = \frac{x}{12}$$

with x, the distance on the map, then being 2 2/5 centimeters. A 13-year-old, for example, measures everything right away and decides to draw everything to quarter scale.

Also, not until Stage 4 will children use the conventional horizontal and vertical or one set of coordinate axes on the map as a basis for locating objects relative to each other.

Until Stage 4, children use a network of horizontal and vertical lines as a reference system.

---

[3]Jean Piaget and Barbel Inhelder, *The Child's Conception of Space* (New York: W. W. Norton & Company, Inc., 1967), p. 444.

Teaching Implications

This activity reveals a great deal about the extent to which children understand space and use mathematics to solve spatial problems. It should be of particular interest to social studies teachers in gaining an awareness of children's ability to consider such geographical concepts as distance and direction.

The construction of a diagram or plan necessitates the selection of a reference system, such as intersecting axes—most often north-south and east-west. These axes are usually related to right-left and up-down on the map to determine angle or direction. Then, to determine distance, proportions must also be considered.

The social studies of geography, economics, and history all involve fundamental mathematical ideas. Time concepts, not attained by many young children, are involved in studying history. (Activities 4–12 through 4–17 explore concepts of time.)

PERSPECTIVE

Purpose of Activity

The purpose of this activity is to study a child's ability to predict what an object will look like from different points of view. To do so he must be able to imagine straight lines (of sight) facing in any direction. This is the essential requirement for forming **perspective** or **different points of view.**

Materials Needed

1. A pencil or stick.
2. A thin metal disk or coin.
3. A doll.
4. Two pictures of railroad tracks going into the distance; one with the tracks remaining parallel, the other with the tracks converging.

Procedure

Two procedures are used. In the first, place the pencil or stick in front of the child and place the doll at right angles to him so that the doll's view of the object is 90 degrees different from that of the child.

 Doll

Child

With the pencil positioned as in the illustration above, the doll, of course, has an end view, which is a circle. Ask the child to draw what the pencil looks like to the doll or to pick from a collection of drawings the view the doll sees.

Variations of this activity involve placing the stick on the table in a vertical position before the child and gradually moving the upper end away from him, asking him to predict how the stick will look as the process is continued. (From the standpoint of number, does he think that the length changes?)

This activity with the pencil or stick is then repeated using a flat disk or coin. When this appears as a circle to the child it should appear as a line segment or thin rectangle to the doll.

The second procedure in studying the child's understanding of perspective and distance involves railroad tracks. Ask the child to draw railroad tracks (and ties) going away from him. Or show him two pictures of railroad tracks, one set converging and one remaining parallel, and ask him to pick which drawing is correct.

Levels of Performance

Stage 1. There is no understanding of the idea of perspective.

Stage 2. From 4 to 6 or 7 years of age there is still total or partial failure. First the object is drawn the same regardless of the point of view. The end-on view is particularly difficult. There is some success with the converging railroad tracks.

At this stage the children realize the picture is different, but they can go no further than what they see. They are perceptually bound and unable to represent or imagine how the object looks from different viewpoints.

Stage 3. Around 7 to 8 years of age, clear distinctions are made between different points of view, at first in general shape and finally in terms of the degree of change in shape. At 8½ to 9, coordination is more perfect and perspective can be applied systematically to drawings.

Stage 4. Usually not until 11 or 12 years of age are children operational with regard to more complex figures such as single and double cones (△, ◁▷).

Teaching Implications

It is surprising that young children have only one viewpoint, their own, as far as the appearance of objects is concerned. Furthermore, they consider everybody's viewpoint to be the same as their own.

There has been little attention to projective geometry in the elementary school, so stages of development have not been a problem. Stage 3 children, ages 8 to 11, might, however, enjoy and profit by activities similar to those just described, particularly if they can check their predictions by moving around the object being considered.

With a light source and screen, shadow projection can also be used to study projective geometry ideas.

SUBDIVIDING A LINE

Purpose of Activity

To begin a study of space, one may logically start at a certain position or point and then consider other geometric ideas such as lines or triangles as sets of points that meet certain conditions. But from the psychological standpoint, can a child accept or envision a logical definition of a line as a set of points? Some materials in geometry today were apparently written on the assumption that the child can, but this activity indicates the difficulties involved.

Materials Needed

Paper and a pencil.

Procedure

Show the child a line segment drawn on a piece of paper. Then ask him to draw another line segment half as long, then another half as long as that. When he cannot draw one any shorter, ask him to continue the process in his mind: "What would be left before there is nothing at all?" "Would it have any shape?"

Levels of Performance

Stages 1 and 2. Children under the age of 7 or 8 have great difficulty in arranging a series of line segments in descending order of length or in understanding what is meant by half of one-half. Their smallest line segment is of a perceptible size.

Stage 3. From 7 or 8 to 11 or 12 years of age, the child can make a series of successively shorter line segments and realize the possibility of a large number of subdivisions. He does not, however, regard the number of subdivisions as infinite.

Stage 4. Beginning at around 11 to 12 years of age, thought is no longer bound by what the child "sees" or by concrete operations based on drawing the line segments. The child now envisions the line as being subdivided into an unlimited number of points, with the points having no shape or surface.

Teaching Implications

The idea of *unlimited* or *infinite* is a complex notion and, in terms of teaching geometry, a notion children are not ready for in the elementary school. To begin, then, with definitions of geometric ideas such as triangles or lines as sets of points that meet certain conditions is to ignore the abilities of the children. Line segments should be considered as line segments—not as sets of points. Hence, the children may, for example, arrive at their own definition of a triangle as being a closed figure consisting of three line segments.

RECIPROCAL IMPLICATION AND EQUAL ANGLES

Purpose of Activity

The purpose is to determine when children can use the logic of **reciprocal implication** and **equal angles** to solve the problem of bouncing a ball off a wall in order to hit an object.

Materials Needed

1. A ball.
2. Blocks or some other target object.

Procedure

Show the child a ball on a table or floor and ask if he can roll it against a wall so that the ball will roll away from the wall to hit another object. If a spring load mechanism to fire the ball is available, so much the better.

Levels of Performance

Stage 1.   These children work empirically, by trial and error, to make a hit but do not analyze the process. They consider mainly the starting point and the target; they are not concerned with the point on the wall that they must hit.

Stage 2.   At the concrete operational level, children begin to relate the point of rebound with the path the ball takes after the rebound, but they are more concerned with the point of contact than the equal angles involved. They are beginning to see that a certain action implies a certain result (implication).

Stage 3.   At this stage, children are successful in analyzing the process and describing it, but this does not usually occur until 11 or 12 years of age—the formal operational level. The child observes

The child attempts to roll the ball so that it will strike the board and rebound to strike the blocks.

that the ball bounces off the wall at the same angle it strikes the wall. (The angle of incidence is equal to the angle of reflection.) The reciprocal implication is to realize that to get the right rebound, the ball must strike the wall at the same angle.

Teaching Implications

The ability to realize the implication of an action and the reciprocal of that action is a logical process. Here it is applied in a framework of Euclidean geometry and physics. Sime concludes the ability to use this process as a tool is a necessary condition to studying Euclidean geometry.[4]

---

[4]Mary Sime, *A Child's Eye View* (New York: Harper & Row, Publishers, 1973), p. 70.

4

Measurement

It is not difficult to see how the facts brought to light in the present study of the psychogenesis of metrical notions may some day be given a practical application in teaching.[1]

So long as the subject's [child's] conception of space is egocentric, it remains a law unto itself, because it lacks references which are coordinated, and changes of position which are grouped . . . measurement is impossible.[2]

The growth of knowledge is not a matter of mere accumulation, and while it is true that between the ages of 4 and 10, children collect a good deal of information about their district, they also coordinate the picture which they have of it, which is an infinitely more complex process of development.[3]

Operations involved in measurement are so concrete that they have their roots in perceptual activity (visual estimates of size, etc.) and at the same time so complex that they are not fully elaborated until some time between the ages of 8 and 11.[4]

---

[1]Jean Piaget, Barbel Inhelder, and Alina Szeminska, *The Child's Conception of Geometry* (New York: Basic Books, Inc., 1966), p. vii.
[2]Ibid., p. 26.
[3]Ibid., p. 24.
[4]Ibid., p. vii.

Measurement will be studied in terms of the following activities:

Length

4-1. Conservation of Length

4-2. Conservation of Distance

4-3. Subdivision and Conservation of Length

4-4. Beginning Linear Measurement

4-5. Measurement of Length

Weight

4-6. Conservation of Weight

4-7. Weighing with a Beam Balance

Area and Volume

4-8. Conservation of Area

4-9. Measurement of Area by Unit Iteration

4-10. Measurement of Area by Multiplication

4-11. Conservation and Measurement of Volume

Time

4-12. Time in Terms of Age—Sequencing Events I

4-13. Time in Terms of Age—Sequencing Events II

4-14. Speed and Time

4-15. Conservation and Measurement of Time

4-16. Synchronism

4-17. Using a Proportion to Compare Speed, Distance, and Time (See Activity 2-16, Ratio and Proportion)

CONSERVATION OF LENGTH

Purpose of Activity

Measuring the length of an object would be difficult if not impossible if the length of the ruler changed as it was moved along the object to be measured. Yet many children think the length of an object changes as you move it, as this activity points out.

The Teaching Implications at the end of this activity summarize the prerequisites for meaningful linear measurement.

Materials Needed

Two ice-cream sticks.

Procedure

1. Show the child the two sticks positioned as shown:

   Ask the child if the sticks are the same length or if one is longer. If she thinks one is longer, place the sticks closer together until she is convinced they are the same length.

2. Then move one stick up or down

   and ask the child if the sticks are still the same length or if one is longer. "Which one? Why?"

   If you think the wrong answer was prompted by the words used, ask "If an ant began at one end of each stick and walked to the other end, would one ant walk farther?"

Levels of Performance

Stages 1 and 2. Between 4½ and 7 years of age, most children think the length of an object changes relative to another object when it is moved. The child will say that the stick that is moved up is longer because she looks at the end farther away and thinks the stick has stretched. Her judgment is made on a perceptual basis and her level of mathematical thinking is topological.

Toward the end of Stage 2, the child may put the stick back where it was, find it again the same length as the other, and conclude that she must have been wrong. Her approach is by trial and error, sometimes right, sometimes wrong, depending on such perceptual factors as how far the sticks are apart or at what angle they are to each other.

Stage 3. Usually soon after 7 years of age the child responds immediately that the sticks remain the same length. Her answer is now based on logic rather than perception. If asked to justify her answer, she uses the logic of identity: they were the same length so they are still the same length. To prove it, she will put the stick back where it was. This is the stage of **conservation of length.**

Teaching Implications

This and the following four activities relate to basic psychological difficulties in learning to measure length. Regarding this activity on conservation of length, the question should be asked, How can the child measure if her ruler or measuring unit changes in length (to her) as she moves it along the object she is measuring?

The next activity reveals another difficulty; i.e., children under 6 or 7 years of age think that the distance between two objects changes if another object is placed between them. They also think that if one object is higher than another, it is farther from the lower to the higher than from the higher to the lower. They have not developed the idea of the symmetric property.

Activity 4-4 involves the transitive property as a thought process. To compare two lengths or heights, such as those of Towers A and C with a ruler, B, the child must realize that if Tower A equals B and Tower C equals B, then Tower A must equal Tower C.

Each of these activities reveals prerequisite thought processes for meaningful measurement that we do not find in many children under 7 years of age. There is still a time lag of about one year (between the ages of 8 and 9) before these mental processes are coordinated into a meaningful whole for measurement. The reason for the time lag is that the activities described here involve thinking only in qualitative terms about physical objects—for instance, which object is longer.

CONSERVATION OF DISTANCE

Purpose of Activity

No linear measurement is possible until the child can realize that the distance between two objects remains constant when another object is placed between them (such as a measuring ruler, for example). Such a realization is not simple for children, as this activity points out.

The idea of the symmetric property, often taken for granted, is also necessary for measurement; that is, the distance from A to B is the same as the distance from B to A.

Materials Needed

1. Two figures such as dolls or trees, the same size, and also one twice as large.
2. Bricks or wooden blocks.
3. A cardboard screen.

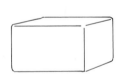

Procedure

1. Place the two dolls (same size) on a table about 50 centimeters apart. Ask, "Are the dolls near or far apart?" Then place the cardboard screen (held vertically) midway between the dolls, and ask the child if they are still near or far apart—to see if she thinks the distance has changed. Also ask why she thinks so. Then substitute another object such as a brick or a block for the screen and repeat the questions.

2. A second procedure involves different sized dolls to see if the child thinks the distance from A to B is the same as from

B to A. Or one of the dolls of the same size can be placed on top of a block.

Levels of Performance

Stage 1.   Up to 4 or 5 years of age, the child cannot consider the distance from A to B as a simple distance when a screen is placed between them. The screen, for her, ends the distance relationship between A and B. She now only considers the distance from one object to the screen and says it is shorter. Being egocentric in outlook, she may also consider only the distance from herself to the object rather than the distance from one object to another.

Stage 2.   From 5 to 7 years of age, the child can consider the idea of distance between objects even when a screen or other object is placed between them. But she thinks the distance is now shorter since the screen has taken up some of the distance or space between the objects. She does not, then, conserve distance.

Also, the distance relation is not symmetrical for her. When one of the dolls is larger, or is placed on a block, she does not think the distance from A to B is the same as from B to A.

Stage 3.   The necessary logical "groupings" in the mind of the child for a logical solution to these problems of conservation of distance and the symmetric relation occur around 7 years of age. Some children between 6 and 7, however, may solve the problem of the screen placed between the objects.

Teaching Implications

The lack of readiness to solve the problem of conservation of distance is based on an inability to think of space without the dolls and of two points in *space* representing the positions *occupied* by the dolls. The gap or distance between these two points is not, of course, affected by other points or objects placed between them. But for children under 7, attention is on the dolls, which seem to draw closer together as objects are placed between them.

The preoperational child makes decisions on the basis of perceptual factors. It looks farther going up (from the small doll to the big doll) than vice versa. Or it "looks like" the objects between the dolls have taken out some of the space between them. The child at the preoperational level is unable to separate the objects, themselves, from the positions they occupy in space as a basis for a logical rather than perceptual solution to the problem.

That the distance from A to B is the same as the distance from B to A (the symmetric property) also involves the idea of reversible thought. Children often do not have this idea until around 7 years of age. The teacher, then, must ask herself how a Stage 1 or Stage 2 child can be taught to use a ruler to measure some distance in any meaningful fashion. The distance relation, for this child, is not a constant quantity, as objects (such as a ruler) are placed between the objects A and B in the course of "measuring."

The teacher must use activities such as those just outlined to determine which children are ready for measurement. The next activity is also necessary to ascertain measurement readiness because many children under 7 years of age think the *length* of an object itself (such as a ruler) changes as it is moved.

SUBDIVISION AND CONSERVATION OF LENGTH

Purpose of Activity

In the preceding activities, preoperational children were found to be unable to understand linear measurement. They think the length of the measuring instrument changes as it is moved in measuring some object.

What about the length of the object being measured? Does it stay the same length when divided into smaller subunits for measurement? Many children think not, even though this is a basic premise in measuring length.

Piaget describes linear measurement as a synthesis of the operations of **change of position** (of the measuring instrument or ruler—the subject of Activity 4-1) and **subdivision** (of the object being measured—the subject of this activity).

Materials Needed

Two strips of paper the same length—approximately 15 centimeters.

Procedure

Show the children the two strips:
"Are the strips the same length?"
"Yes."
"How about if we cut one and arrange it like this?

125

Or like this?

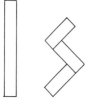

Are they still the same length or is one longer? Why?''

Levels of Performance

Stage 1.   Children 5 to 6 years of age consider the straight, uncut piece to be longer.

Stage 2.   At 6 to 7 years, children oscillate between correct and incorrect answers, working out the problem by trial and error. They may straighten out the zigzag path and find it to be the same length as the straight path.

Stage 3.   Children 7 to 8½ years of age give correct responses immediately.

Teaching Implications

This and the preceding activity explore the important conservation concepts necessary for understanding linear measurement. In terms of *understanding,* most children are not ready for linear measurement until about the third-grade level.

According to Piaget:

> The idea of necessary conservation [in order to measure], which entails the complete coordination of operations of subdivision and order or change in position, is accomplished at stage III. This stage was found to have been reached by one in ten in the age range 6–7, by half of those 7–7½, and three-quarters of those 7½–8½ years old.[5]

---

[5]Jean Piaget, Barbel Inhelder, and Alina Szeminska, *The Child's Conception of Geometry* (New York: Basic Books, Inc., 1960), p. 114.

BEGINNING LINEAR MEASUREMENT

Purpose of Activity

Measuring length with a ruler seems so simple to the person who knows how. It is difficult to realize the complexity of the problem for the child who does not. This activity helps identify the stages through which children go during their first attempts at linear measurement.

Materials Needed

1. Two tables or two boxes of different heights.
2. A screen to separate the tables.
3. Two sets of wooden blocks with which to build towers (the blocks should be in different sizes and shapes so the child cannot match blocks in one tower with those in the other).
4. Sticks longer and shorter than the towers.

Procedure

Show the child a tower of blocks standing on one table (about 50 centimeters high). Ask her to build another tower on the other table "the same height as mine." Place a screen between the tables so the child cannot compare the two towers simultaneously, but allow her to go and look at the model tower as often as she wishes. The wooden blocks are different sizes so she cannot count or match blocks in the two towers. The tables are different heights.

Levels of Performance

Stage 1. From 3 to 4½ years, comparison of the two towers is perceptual. The child builds a tower, looks at the model and says they are the same height. "Why?" "Because I looked." Given a stick and asked if that would help, she says, "No," or puts it on top of the tower as a decoration.

Stage 2. During the period from 4½ to 7 years of age, the child still begins on a perceptual or visual basis, sighting from one tower to the other, connecting their tops with one of the sticks (ignoring the bottoms), and so on. This method might be called **visual transfer.**

Toward the end of Stage 2, the child begins to use a common term—her body, or her hands, for instance. Putting one hand at the top of the tower and one at the bottom, she then walks to the other tower with her hands held the same distance apart to see if this length fits the other tower. This method of **manual transfer** follows the visual transfer or "looking" method. Manual transfer involves imitating the towers with her body.

The use of the body as a third term to compare to the other two terms (towers) is followed by use of the sticks, which is a move toward the abstract or symbolic. The stick must, however, be the same length as the towers.

Stage 3.   Between 7 and 9 years of age, the idea of an independent third term such as a stick or ruler to compare the two towers can be used. The thought process necessary for this performance is the concept of the **transitive property.** If the stick, A, is equal to or the same length as one tower, B, and also the same length as the other tower, C, then B must be the same length as C.

During the first part of this stage, only sticks as long as or longer

A 6-year-old attempts to build his tower to the same height as the model towers on a higher table.

than the towers can be used as a third term. During the latter part of the stage, sticks shorter than the towers can be used as a measuring instrument. The child moves the stick along the tower, marking and counting. This, of course, is true measurement and what Piaget calls **unit iteration.**

The idea of the transitive property, which involves relating three terms and making the necessary comparisons, calls for logical or operational thought. Also necessary is a spatial reference system allowing the child to coordinate the ideas of two towers and a stick and make allowances for the different heights of the tables.

Teaching Implications

This activity indicates the level at which a child thinks in beginning measurement—a level which may very well be below what the teacher thinks the child can do. Piaget remarks that performance in Stage 2 may be complicated by what the child has "learned" in school, since what she has been shown or told about measuring cannot be truly understood until she has the necessary mental structures outlined in Stage 3.

Even if there are not two towers to be compared but only one object to measure, the child must first reach the stage of conservation of length and also be able to consider the idea of a whole divided into parts for measurement purposes.

MEASUREMENT OF LENGTH

Purpose of Activity

To measure an object with a ruler or other unit of measure, then measure another object, and finally compare the measurements obtained from the two operations in order to say which object is longer involves truly metrical, or quantitative, as well as qualitative thinking.

According to Piaget, the necessary quantitative operation for true measurement takes place in the mind of the child by 8 to 9 years of age. He is convinced that the child arrives at this stage not through adult intervention, but largely on her own.

Materials Needed

1. Strips of paper.
2. Cardboard.
3. Glue or paste.

Procedure

To investigate the child's readiness (at 8 to 9 years in most cases) for true linear measurement, paste strips of paper on cardboard in various configurations such as:

Then give the child a cardboard strip the same length or shorter than the segments of the figures to be measured and ask her to find which figures are longest, shortest, and so forth. She may be shown how to start by placing the cardboard strip at one end of a figure to be measured, marking its end-point, and then moving the strip to the next position, calling these moves "steps."

Levels of Performance

Stage 1. Children at this stage have not constructed a single spatial framework that involves the space occupied by the objects as well as the objects. They are not able to coordinate the ideas of change of position and subdivision explored in the two preceding activities.

For example, a child, 6½ years of age, who is asked to compare two shapes of equal length ∟ ∠ measures A with a smaller
$\qquad$ A   B
strip and says 6, and then says it is longer than the other (B) without even measuring.

Stage 2. These children are moving toward conservation but have not fully arrived. They are beginning to see the role of a measuring unit but they need prompting. For example, one child slides her measuring strip (ruler) aimlessly along the object being measured:

"Can you tell that way?"

"No."

"Then what can you do?"

She then corrects her procedure and concludes that one object is longer because she gets 6 for it and 5 for the other.

Given a more difficult figure to measure, such as a zigzag staircase with some stairs shorter than the strip of paper with which she is measuring, she ignores the difficulty and puts in measuring marks arbitrarily.

Stage 3. Children 8 to 9 years of age are operational for linear measurement. They can select and use an appropriate measuring unit and compare lengths of various objects successfully. They do not use the trial-and-error approach of Stage 2. Instead, they develop an appropriate plan of attack and immediately begin to employ a correct procedure to solve the problem. Any errors made are quickly corrected.

Teaching Implications

This activity provides procedures and materials children can use to reinforce their measuring techniques once they have the necessary developmental characteristics involved in the preceding three activities. Now the standard units in the metric system can be explored and compared, and appropriate units selected; e.g., the centimeter for small objects and the meter for larger ones.

CONSERVATION OF WEIGHT

Purpose of Activity

The purpose of this activity is to explore children's understanding of the idea of weight.

Materials Needed

1. Two balls of plasticene or clay the same size.
2. A two-pan balance.

Procedure

Give the child the two balls and ask if they weigh the same. If the child thinks that one is heavier, ask her to pinch some off until she is satisfied that the two balls weigh the same. Then flatten one ball like a pancake or roll it into a hotdog shape and again ask the child if one is now heavier or if they still weigh the same. Finally, ask why.

For the child who says the weight has changed, place the two pieces of clay on a two-pan balance to see if that causes her to change her answer. (As will be seen in the next activity, understanding the weighing process is more complicated than may be expected.)

Levels of Performance

Stage 1. These children can give no reasonable answer. They think the weight changes as the shape changes.

Stage 2. At this stage, children may at first think the weight changes, but the act of weighing changes their minds for the weighing in question. They may be unable to generalize to another similar experiment.

Stage 3. At approximately 8 years of age, children are concrete operational. They give correct and immediate responses. They realize that changing shape does not change weight. The weight is conserved (remains the same).

Teaching Implications

Weighing cannot be a meaningful activity for children if they think that shape affects weight. Their weighing activities would have to involve objects the same shape.

Also, as will be seen in the next activity, there is the problem of understanding how a balance "works." The idea is not as easy to grasp as many teachers think.

WEIGHING WITH A BEAM BALANCE

Purpose of Activity

The purpose is to explore children's understanding of weighing using a beam balance.

Materials Needed

A beam balance—homemade or commercial.

Two children, 9 and 10, Stage 2, successful only by trial and error.

Procedure and Levels of Performance

1.   Show the child the balance with no weights on it but with an empty cup at each end. Then ask what will happen if a weight is put in one of the cups:
    "It will come down."
    "Then how can you make it level again?"
    "Push it up."

135

The **Stage 1A** child, being egocentric, thinks the balance should respond to her action. She is surprised the arm will not stay up. The **Stage 1B** child realizes that weight can be added to the other arm to make the arms balance (reciprocity), or that the weight first put on can be taken off (negation).

2.   With the balance even and an empty cup at the end of each arm, put 2 weights in one cup and 1 weight in the other:
"Is there anything I can do to make it balance without adding or taking away weights?"
"Move the cup nearer the middle."
"Which one?"
Children at Stage 1B are not sure which cup to move, but by trial and error are successful.
"Now that it is level with 1 weight in the cup at the end and 2 weights in the cup halfway along, what will happen if I put another weight in the cup at the end?"
"It will stay the same because each side has 2 weights."
The weight is added to the end cup and that side drops.
"Now what?"
"Move the cup that is halfway out all the way out."
The beam then balances.
"Now if we add another weight to the left cup, what must we do to make it balance?"
"Move the other cup in."
This action, of course, creates more of an imbalance.
At Stage 1B, children realize that there is a relation of distance as well as weight involved, but they are unable to resolve it. They use the principle of reciprocity, trying to keep the weights equal or the distances equal on both arms, or they use the principle of negation in two ways: by taking off a weight that has been put on or by moving inward a weight that had been moved out (or vice versa).

3.   With the two cups at each end of the balance and 1 weight in each cup, ask "What will happen if I add another weight to the left cup?"
"It will go down."
"How can I make it balance without adding another weight?"
"Move the cup with the two weights in toward the center."
The **Stage 2** child can make the cups balance by moving the heavy cup in toward the center, but she does not know how far. She is successful by trial and error. At **Stage 3,** 11 to 12 years of age, the child is sure of how to solve the problem. She realizes that there is

an inverse relation between weight and distance: the smaller the weight, the farther it must be from the center to balance a heavier weight. She also can use the logic of simple proportion to solve the problem. One weight at the end of the arm would balance 2 weights halfway to the end of the other arm.

The Stage 3 child moves the cup containing 2 weights immediately to the halfway position, knowing she will be right to balance one weight at the end of the other arm. She does not have to use the trial-and-error method of the Stage 2 child to "see" if she is right.

Teaching Implications

Children do a lot of weighing using balances in the primary grades, but they do not understand the process, as this activity reveals. A 6-year-old making a cake that requires equal amounts of flour and sugar does not know to which bowl to add if, for example, the one with sugar is lower. She may be told where to add the sugar, but the telling does not provide the understanding.

Weighing on a beam balance is not simple in terms of the logic processes involved. The logic is that of the INRC (the four group developed by the Bourbaki school of mathematics):

I   identity: all the weights can be removed to restore balance or identity.

N   negation: a weight put on can be taken off.

R   reciprocity: equal weight can be added at the same distance from the fulcrum on the other balance arm.

C   correlation: lighter weight can be added farther away or heavier weight added closer to the fulcrum.

CONSERVATION OF AREA

Purpose of Activity

The purpose of this activity is to determine when children reach the stage of conservation of area as a prerequisite for understanding area.

Materials Needed

1. Two rectangular cardboard sheets painted green, approximately 10 by 12 inches each.
2. Two small wooden cows.
3. Small wooden blocks to represent houses.
4. Two cardboard rectangles the same size and shape, one cut diagonally so its shape can be changed to form a triangle.

The child points to the pasture which he thinks has more grass. (Child at left is Stage 3.)

Procedure

Two procedures are used. In the first the child is shown the two green pieces of cardboard with a wooden cow on each. The child is asked if each cow has the same amount of grass to eat. If she thinks they do, she is then told that a farmer has decided to build a house on each field. A small wooden block is put on each field. Again the child is asked whether each cow has the same amount of grass or green to eat. The process is continued, adding a house to each field, but placing the houses in a row on one field and scattering them about the second.

The second procedure is to show the child two cardboard rectangles and ask if they are the same size or if one is bigger. When it is agreed they are the same size, the rectangle formed by the two triangles is transformed to the shape of a triangle and again the child is asked if they are the same size or if there is the same amount of room.

Levels of Performance

Stages 1 and 2. Up to 5½ or 6 years of age, children solve the problems at the perceptual level. As houses are added but scattered out in one field, the child thinks the cow in that field has less grass to eat.

Toward the end of Stage 2, from 6 to 7 years of age, the child thinks the amount of grass stays the same as several houses are added, but then is overwhelmed by the perceptual configuration as more houses are added; again the grass area must be smaller on the field with scattered houses.

For the second problem, changing one rectangle to a triangle, the child thinks one figure is larger. "Why?" "Because you've cut it," "Because you've turned it around," and so on.

Stage 3. From 6½ to 8 years of age, the child answers the field problem on a logical rather than perceptual basis using the logic of Euclid's axiom: if equals (house areas) are subtracted from equals (field areas), the remainders (grass areas) are equal.

Teaching Implications

The field problem reveals the ability of children to think logically or mathematically concerning conservation of area using the logic of Euclid's axiom.

The second activity, although a study of conservation, involves the mathematical idea of a transformation in shape (or change of position). Changing a rectangular shape to a triangular shape and asking if there is still the same amount of room might be varied in a number of ways to explore the child's level of understanding. The rectangle might be cut to form a thinner, longer rectangle and placed first vertically and then horizontally. Or one end of the rectangle could be cut off and placed on top of the other part.

It is interesting that conservation-of-area problems are solved at about the same time as conservation-of-length problems, since one might think area problems would be more difficult. A full understanding of area measurement does occur later, involving a fourth stage, as will be seen in the next activity.

MEASUREMENT OF AREA BY UNIT ITERATION

Purpose of Activity

The purpose of this activity is to determine stages of development in ability to measure area using as methods

1. superposition and unit iteration; and
2. multiplication of linear measures (length × width).

Materials Needed

1. Cardboard shapes cut out as follows:

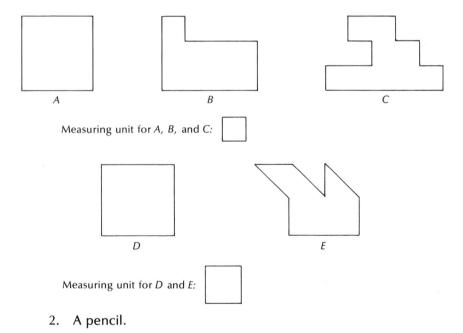

A   B   C

Measuring unit for A, B, and C:

D   E

Measuring unit for D and E:

2. A pencil.

Procedure

The first method of measuring area involves giving the child a small square of cardboard as a measuring unit and a larger square (Figure

A). The smaller square will fit on the larger square 9 times so that its area is 9 square units.

The child is also given a pencil so that she can mark off the area as shown by tracing with her square measuring unit. If the child has difficulty, the interviewer can help by marking off two or three squares as the child watches.

This method of fitting the measuring unit on the object being measured is called **superposition.** Also involved is **unit iteration,** or counting to find how many small squares there are in the larger area; in this example, 9.

The child is asked to compare the size of Figures *A, B,* and *C,* all of which have an area of 9 square units. She is then given a larger measuring unit and asked to compare the size of *D* and *E, E* having an area of 3½ square units as compared to 4 for *D.* In addition to the square to be used as a measuring unit, the child is given a triangle half the size of the square. By using the triangular measuring unit, the top part of Figure *E* is found to have an area of 3 triangular units, so that its total area is one-half square unit less than Figure *D.*

Levels of Performance

Stages 1 and 2. From 5 to 6 or 7 years of age, children are unable to solve the problem for the following reasons:

1. Not having the concept of **conservation of area,** they think the area must change as the shape changes.

2. The idea of moving the square measuring unit 9 times to make a fit with the larger square is meaningless for them. **Unit iteration,** or **additive composition,** is not yet present.

3.   The concept of **transitivity** is also lacking. To compare two figures such as B and C by measuring each with a third term such as the measuring unit requires the logic of transitivity. If 9 measuring units are the same as Figure B and Figure C, then Figures B and C must be the same size.

The judgments of these children are perceptual. Their method involves making both shapes into the same form, such as a square, so that the figures "look alike." But these changes in form or trans- formations are not compensatory such as by taking a square unit from one part of the figure and placing it in another. Additions or subtractions to the figure are purely arbitrary.

Toward the end of Stage 2, from 6 to 7 years of age, children are transitional. First they think one figure is larger. But if, when asked to count, they find 9 units for each area, they say they are the same. Likewise, if asked would they prefer one or the other if each figure were chocolates, they have a preference.

Stage 3.   During the first part of Stage 3, ages 6 or 7 to 8, children solve the problem of Figures A, B, and C, but the idea of a metrical or measuring unit is still not fully understood. In comparing D and E, using both the triangle and the square as measuring units, they count the number of triangles as if the triangle were the same size as the square.

Toward the end of Stage 3, 7½ to 9 years, understanding is com- plete; solution is immediate by superposition and unit iteration.

Stage 4.   Not until 11 or 12 years of age, or the formal operational level, can the child graduate, so to speak, from understanding area by dividing the inside into smaller square units and counting, the method of Stage 3, the concrete operational level.

At Stage 4, the child can consider an area in terms of two linear measures and multiply them. This requires formal or abstract thought beyond the Stage 3 level in which the child must lay out or draw and count the subunits.

Teaching Implications

This activity provides a basis for the analysis of children's ability to measure area and, consequently, a basis for determining when measurement of area should be introduced. Most children 8 or 9 years of age should be ready for measuring area by dividing it up into smaller square units (the method of superposition and unit iteration described in this activity).

Determining area by this method involves three characteristics of logical thought: conservation of area, additive composition, and transitivity. The teacher, of course, should be familiar with them and use them as a basis for diagnosis and lesson planning.

Piaget did not find children ready for the more sophisticated method of understanding area—using linear measures of length and width and multiplying—until they reached 11 or 12 years of age, the level of formal operational thought. This is the subject of the next activity.

MEASUREMENT OF AREA BY MULTIPLICATION

Purpose of Activity

The purpose is to determine when children are able to develop the concept of multiplication of linear measures as a basis for determining area (area = length × width).

Materials Needed

1. Drawing of a line segment 3 centimeters long.
2. Drawing of a square measuring 3 centimeters by 3 centimeters.
3. Paper and a pencil.

Procedure

First show the child the line segment and ask her to draw one twice as long. Then show her the square and request that she draw a square (not a rectangle) twice as large (18 square centimeters).

Levels of Performance

Stage 2. At 5 to 7 years of age, children increase the size arbitrarily, having no basis for calculation.

Stage 3A. At 8 to 10 years of age, children do one of two things: they either (a) double the area of the square by making another one like it, thus ending up with a rectangle rather than a larger square, or (b) double the length of the sides of the square, which results in a square four times as large as the model.

Stage 3B. The child starts as in Stage 3A by doubling the length of the sides of the square but then realizes that the new square will be too large. She then tries to establish some other relation between the length of the sides and the new area, but is unable to do so.

Stage 4. At this stage, 11 to 12 years of age, the formal operational level, the child is finally able to approximate the relation between length of sides and area. She knows that she must find a number for the length of the side which, multiplied by itself, will be

18. She may try 6, find it too large; then 4, find it too small and say that it is close to 4.

A 12-year-old reasons as follows:

"This one is 9 square centimeters. I need one twice that, or 18 square centimeters. I divide by 4, since there are 4 sides which gives me 4.5 centimeters for each side, but that's still not quite right."

"Why not?"

"That's still too large—4.5 by 4.5. You can't work it out."

This child is close to the solution, even though she does not know how to calculate the square root of 18 as a basis for determining the length of the sides.

Teaching Implications

While the answers at Stage 4, 11 to 12 years of age, are still not correct, these children do realize that the area of a square can be determined by multiplying the length of a side by itself. Their answers at this age are partially influenced by the schooling they have had, such as having been exposed to the formula $A = l \times w$.

At this level, the child can think abstractly of the unlimited number of subdivisions that can be made in a given length and width, providing a rationale for multiplying length and width to determine an area. It was, of course, found in an earlier activity on linear measurement that not until 11 or 12 years of age were children able to think of a given length as an unlimited set of points.

If Piaget is correct, when we "teach" children a formula, $A = l \times w$, before they are 11 or 12 years old, utilizing this formula can only be a rote procedure for many of them.

CONSERVATION AND MEASUREMENT OF VOLUME

Purpose of Activity

The purpose of this activity is to determine when children are able to understand conservation and measurement of volume.

Materials Needed

1. Thirty-six small wooden cubes, each approximately 1 cubic inch in size.

2. A wooden block having the same volume as the thirty-six small cubes and the form of a rectangular solid (base 3 × 3 and height 4).

3. Cardboard rectangular shapes to serve as islands on which houses can be built with the small wooden cubes. Bases for the houses (or islands) should have each of the following dimensions: 2 × 3, 1 × 2, 3 × 4, 1 × 1.

4. A bowl of water large enough to submerge the thirty-six unit cubes in the form of houses in various shapes.

Procedure

The general plan is to show the child the wooden block and tell her that it is a house on an island. Since a storm is expected she will be asked to build another house with the same amount of room, but on a smaller island. She will then be given the thirty-six unit cubes and a cardboard island on which to build the house. Possibilities include those in the illustration on the following page.

As a prelude to the questioning, first ask the child to make a house like the wooden block (the same size and shape). She may be helped if necessary. Then take these blocks and make another house with a different-sized base and ask the child if the new house has just as much room, more room, or less room than the old house. Also ask why she thinks so. This, of course, is one test of conservation of volume.

147

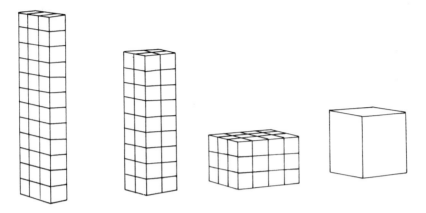

Then give the child one of the cardboard rectangular shapes representing an island and ask her to build a house on it that has the same amount of room as the house just built.

Levels of Performance

Stage 1.   The technique is impractical for children under 4 or 5 years of age.

Stage 2.   From 4½ or 5 years of age to 6½ or 7, children at first refuse to build a taller building to allow for a decrease in the size of the base. They think in one dimension only. They start to build the house higher, but think that if it is higher it must have more room. They cannot conserve volume.

Toward the end of Stage 2, children begin to compensate for the smaller base by building a structure that is taller—but not tall enough. Such children are beginning to establish the relation of volume to the three dimensions of length, height, and width.

Stage 3.   This stage begins around 6½ or 7 years of age with the child's making modifications in height and width and length to adjust for the different-sized base. She has reached a stage of conservation of volume but has not perfected a method for building another house with the same amount of room. To do this, she will have to count the blocks and multiply the three dimensions—a metrical solution.

Toward the end of Stage 3, 8 to 10 years of age, the child uses a metrical method, counting the blocks in each dimension and knowing that the length times the width times the height (in blocks) should be 36. She then can make the necessary adjustments in her new house. If the building has a base 2 × 3, then its height must be 6 if the volume is to be 36. If the building has a base of 1 cube, then it must have 36 stories.

Stage 4.   Not until 11 or 12 years of age, or the formal operational level, does the child understand a second aspect of conservation of volume. In Stage 3, the child realizes the room "inside" the blocks or "inside" the house does not change as the shape of the house is changed. This is **conservation of interior volume.** But if two houses of different shape are immersed in water, the Stage 3 child does not think that each will make the water rise the same amount. This concept that the space "occupied" by the house is also invariant is a consideration of space in a wider context and a stage not reached until the child is almost through the elementary school.

Teaching Implications

Although this activity is a diagnostic one, it also provides many opportunities for children to explore the space relations involved in volume considerations. The procedure offers a way of determining to what extent children understand the formula for volume which they are usually given; that is, $V = l \times w \times h$.

This study would indicate that Stage 3 children are ready for volume measurement at around 10 years of age, although their understanding of occupied space is not complete for another year or so.

TIME IN TERMS OF AGE— SEQUENCING EVENTS I

Purpose of Activity

The purpose of this activity is to determine what basis children use for determining or comparing ages. Can children place things in temporal order or sequence events in time? For example, if A is older than B, who was born first?

Materials Needed

None.

Procedure

Ask the child of 4 to 9 years of age if he has any sisters or brothers and what their names are. If he has no sisters or brothers, ask if he has a friend. Let us assume that the child being interviewed is 5 years old and that he has a sister, Sue, 7 years old:

"How old is Sue?" "Don't know."
"Who is older, you or Sue?" "Sue."
"Why?" "Because she's bigger."
"Who was born first?" "Don't know."
"How old were you when you were born?" "Don't know."
"Who will be older when you start to school?" "Sue."
"Who will be older when you are grown up?" "Me."
"Why?" "I'll be bigger."
"Is your mother older than you?" "Yes."
"Is your granny older than your mother?" "No."
"Why?" "Mother is bigger."
"Does your granny grow older every year?" "No, she stays the same."
"How old was granny when she was born?" "She was old right away."

Levels of Performance

Stage 1. The child does not grasp the idea of time as a continuum. Age is determined by size. He thinks that aging stops when you are grown up, therefore mother and granny are the same age. He may think that he is older than his parents because he was here when he first saw them. He cannot reason that because Sue is older he must have been born later. He cannot sequence events in time.

Stage 2. From 6 to 8 years of age, the child can reason who was born first based on present ages, but he thinks that as he grows he will become older than Sue because he will be bigger. Age (time) is equated with size (space) and is not a continuum in its own right.

A second type of child at this stage cannot tell who was born first based on present ages but realizes the age differences remain constant as he grows older. As one child put it, "People grow bigger, and then for a long time they remain the same, and then quite suddenly they become old."[6]

These children have an intuitive or perceptual understanding of time. Toward the end of Stage 2, if questioned properly, they arrive at the correct answers on a trial-and-error basis.

Stage 3. Around 8 years of age, children's answers are usually immediate and correct. Their answers are now on a logical rather than perceptual basis. These children (at the concrete operational level) understand

1. the idea of *succession of events in time* (in terms of order of births)—"If I am younger, I must have been born later."
2. the idea of *duration or conservation of age differences*—"If Sue is 2 years older, she will always be 2 years older."

Teaching Implications

Children's answers to age questions are first based on size. Whoever is bigger is older. Thus, time as measured by age is, in turn, measured by the spatial idea of size. Time (temporal succession) and space (spatial succession) are not separated as ideas and will not be until around 8 years of age.

There are two particularly important implications. First, children hear parents talk about their ages, their "younger" brother, and so on, but the necessary logical or mental operations for understanding relative ages and time are not present until around the age of 8. Instruction in the conventional sense, by verbal explanation, will not

---

[6]Jean Piaget, *The Child's Conception of Time* (New York: Basic Books, Inc., 1969), p. 208.

convey true understanding, but only a correct response at the verbal level. Thus,

"Joe was born first."

"How do you know?"

"Because you told me."

Second, if the young child of 4 to 6 years thinks he is older than his parents because he was there when he first saw them, then where is the logic of "mother knows best"? Mother cannot give as a reason that she knows best because she is older and, therefore, wiser. If the child, however, equates age with size, then mother is older because she is larger. The logic then, of course, is larger → older → wiser.

Another problem is the egocentric nature of young children. It is difficult for them to look at a situation from other than their own point of view. In a family of three brothers, for example, the child often says there are only two, since "brothers" relates only to him.

Emotional insecurity may also affect the child's answers. While he may be older now than a younger brother, he may not feel that he will grow up to be a man as quickly.

TIME IN TERMS OF AGE— SEQUENCING EVENTS II

Purpose of Activity

This activity provides another approach to study children's concepts of age.

Materials Needed

1. Pictures of six apple trees, each tree a different size.
2. Pictures of five orange trees, each tree a different size.

Procedure

Show the child the pictures of the six apple trees. Say, "These are pictures of the same apple tree taken a year apart. The smallest tree is one year old." Ask her to arrange the trees in a row in order of age. Then show her cutouts of five orange trees, a year apart in age.

Put the smallest orange tree above the 2-year-old apple tree, saying
that when the apple tree was 2 years old, a small orange tree was
planted. Then ask the child to complete the row of orange trees
(lining them up with the apple trees). The orange trees are drawn so
that they appear to have grown faster—the last is bigger than the
apple tree under it. Finally, the child is asked,

1. "Which tree is older this year?" (Point at the last tree.)
2. "Why?"
3. "By how many years is it older?"

Levels of Performance

The developmental stages will be found to occur at approximately
the same time as in the preceding activity. Preoperational children
will say the last orange tree is older than the corresponding apple
tree (because it is larger), even though it was planted later than the
apple tree. Again age is equated with size. If it is any solace, little
children think of you as a teacher as younger than your students
who are taller than you.

Teaching Implications

See Teaching Implications for the preceding activity.

SPEED AND TIME

Purpose of Activity

Because time is measured by a constant speed of some measuring instrument such as a clock hand, we need to know what a child's concept of speed is.

Materials Needed

1. Two pieces of cardboard, folded to make tunnels, with one tunnel longer than the other.
2. Two toy cars with strings attached to pull them through the tunnels.

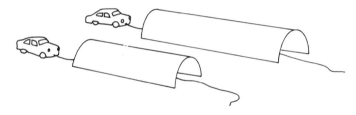

Procedure

Place the two cars at the same end of the two tunnels and make the cars enter and exit from the tunnels at the same time. You may need to make marks on the table to know how far each string must be pulled for the cars to exit at the same time.

First ask the child if one tunnel is longer. Set the cars in motion, entering and exiting from the two tunnels at the same time. Ask the child if the two cars went at the same speed or if one went faster and why.

Then remove the tunnels and mark the paths under them with chalk. Repeat the experiment, moving the cars along the chalk line while the child watches. Ask the child if one car went faster or if they went at the same speed. Then place the tunnels back over the chalk lines and repeat the experiment.

Levels of Performance

Stage 1.   At Stage 1, 4 to 6 years of age, the children agree that the cars went in and came out of the tunnels at the same time, but they think that they went at the same speed because they came out at the same time. Their reasoning is based on the logic of order—out at the same time, hence, same speed.

When the tunnels are removed, these children agree that the car in the longer tunnel went faster, but the reasoning is based simply on one car overtaking or passing the other.

When the tunnels are used again, these children revert back to their first answer—that the speeds are the same because the cars come out at the same time.

Stage 2.   From 6 to 7 years of age, answers are at first the same as those of Stage 1, but the child is able to correct her answers if questioned about length of the tunnels and times of entry and exit.

Stage 3.   At around 7 years of age, the child is able to coordinate the idea of time, distance, and speed. If the cars entered and came out at the same time and one went a greater distance, then it must have gone faster.

Teaching Implications

Piaget finds the notion of speed a more primitive or fundamental one than the notion of time.[7] (The notion of time involves a coordination of speeds—the subject of the next activity.) The child can, therefore, solve some questions concerning speed before she can those concerning time. She can say that one car is going faster because it overtakes or passes another. She is using the logic of **temporal order** in a succession of events. First, A is behind B, then even, then in front of B. Thus, there are primitive or intuitive notions of speed based on the idea of order of events in time, even at 4 to 6 years of age.

There is a second temporal notion which that child cannot understand; i.e., **duration** or elapsed time between two events, such as between entry and exit from the tunnel. Even though the child agrees that the cars enter and exit from the two tunnels at the same time, she thinks that the car was in the longer tunnel a longer time, as the next activity will demonstrate.

There is no primitive notion of time, and an intellectual construction by children is not made until 9 or 10 years of age.

---

[7]Jean Piaget, *Genetic Epistemology* (New York: Columbia University Press, 1970), p. 61.

CONSERVATION AND MEASUREMENT OF TIME

Purpose of Activity

The purpose of this activity is to demonstrate that telling time requires an intellectual construction not usually found in children until 9 or 10 years of age. The preoperational child judges the amount of time based on how much was done or how fast it was done without the necessary condition of relating the two.

Materials Needed

1. Two dolls or toy cars.
2. A three-minute sand timer.
3. A stopwatch.

Procedure

Place the two dolls on a table side by side and say they are going for a walk. Tell the child to say when the dolls should "go" and when they should "stop." When the child says "go," hop the dolls along the table keeping them side by side. When the child says "stop," ask her if the dolls started and stopped at the same time.

In a second procedure the dolls are started and stopped at the same time as before, but the interviewer makes one doll take longer hops so that when it has stopped it has gone farther. The child is then asked again if the dolls started at the same time and stopped at the same time.

Levels of Performance

Stage 1.   Children under 6 or 7 years of age think that the doll that went farther took more time.

To prove that it is not a case of perceptual illusion, ask the child, "When A stopped, was B still going?" and "When B stopped, was A still going?" The child will give a negative response.

Again ask the question, "Then did they stop at the same time?" The answer is again, "No."

These children cannot successfully consider the notion of **simul-**

157

taneity—two things happening at the same time—when the motions or speeds are different.

Stage 2. From 6 or 7 to 9 or 10 years of age, children will agree that the dolls started and stopped at the same time, but the period in between—the **time interval** or **duration**—is still not the same. If asked if one doll walked for a longer time than the other, these children will reply that one did (the one who walked farther). The conclusion is that if the doll went farther (more motion or action), then it must have taken more time. The amount of motion (distance) and the speed at which it happened cannot be successfully related. If more happened, it must have taken longer.

Stage 3. From 9 to 10 years of age, children can relate action or motion to the speed at which it occurred and realize that greater speed accounts for greater distance while the amount of time is the same or conserved.

Teaching Implications

If children confuse or are unable to separate the idea of time from that of speed, they are not ready to understand time. (See also Teaching Implications at the end of next activity.)

SYNCHRONISM

Purpose of Activity

Do children realize that different clocks "tell" or measure the same "time"?

Materials Needed

1. A watch and clock, both with second hands.
2. Paper and a pencil.

Trying to find out if it takes more, less, or the same time to put objects in a cup while the second hand on a clock makes one rotation as to do the same thing while the second hand on a watch makes one rotation.

Procedure

Preoperational children think that the second hand and minute hand do not measure the same time because the second hand moves faster and farther than does the minute hand. Similarly, these children reason that it takes longer for the minute hand to move around the clock than for the hour hand to move from 12 to 1 because the minute hand went farther. These children think that going farther or doing more takes more time. Thus, to cook a 3-minute egg takes more time by a stopwatch than by a timer because the watch "goes faster" (or farther).

To verify these ideas, ask the child to make marks at regular intervals on a sheet of paper while she watches the second hand on a watch make one rotation. Then ask her to make the marks again while watching the second hand on a clock make one rotation. Finally, ask if making the marks took more time one time than the other and why. Instead of making marks, counters can be moved from one container to another.

A 3-minute timer can also be compared to the stopwatch, the second hand on the watch making three rotations while the timer empties. Ask the child if one took more time than the other. Piaget reports that using the stopwatch instead of making marks on paper "does not help at all, because these children have no notion of the constancy of the speed of the measuring instrument."[8]

Levels of Performance

Not until 9 or 10 years of age (Stage 3) can children relate action or motion to the speed at which it occurred and realize that greater speed accounts for greater distance while the amount of time is the same or conserved.

Teaching Implications

Although the adult realizes that instruments used to tell time can move at different speeds (hour hand and minute hand) to measure the same time, the preoperational child cannot. This realization does not occur until 9 to 10 years of age.

Piaget defines time as the coordination of movements or speeds (of the measuring instrument and whatever is being timed).[9] Psychologically, the preoperational child cannot synchronize the durations (of time) of two moving bodies.

Concentrated work on learning to tell time, then, should begin at approximately 9 to 10 years of age; not until then do children reach

---

[8]Jean Piaget, *Genetic Epistemology* (New York: Columbia University Press, 1970), p. 72.
[9]Ibid., p. 59.

the concrete operational level for intellectualizing the idea of time. Tasks such as those just described can be used to determine readiness.

Teachers can introduce the idea of time to younger children on a perceptual basis. For example, "When the little hand points to two it is time to go home" or "When the big hand points to six it is storytime." This might be called "reading" a clock or language arts—the recognition of symbols. "Telling time" is an important survival skill, but it is a long way from understanding the underlying mathematical ideas as these activities demonstrate. Verbalizing explanations of what time really is in terms of clocks will not be meaningful until children understand the processes described in these activities.

activity 4-17

USING A PROPORTION TO COMPARE SPEED, DISTANCE, AND TIME

See Activity 2-16 on ratio and proportion.

5
Knowing versus Performing

The recent emphasis in education on "performance," "competency," and "behavioral-based objectives" is in part an attempt to model educational practices after those of business or industry—to make education more "efficient." The premise is, of course, that what is good in industry is good in education. But performance on a machine is not the same as performance resulting from human mental activity. Correct action at a machine may be programmed using a stimulus-response method, which results in a sensorimotor-type knowledge

Piaget envisions mental activity as much more complex than overt external behavior or performance can demonstrate. His *Grasp of Consciousness* (1976) points out very clearly how different "performing" or "knowing how to" is from "knowing" or "understanding the process" the child is performing. There is often a time lag of as much as five or six years between "performing" and "understanding" what is "performed."

If education is to involve developing understanding as contrasted to performing, or "knowing how to," serious attention needs to be given to the research as exemplified in the activities described in this section. Johnny may know how to add without understanding the process. If this is the case, how will he apply his "know how to" where the process or problem has not been identified for him as one necessitating addition?

163

This section includes the following activities:

WALKING ON ALL FOURS[1]

Purpose of Activity

Is a child consciously aware of his actions? For example, does he "know" what he is doing—when he crawls or walks on all fours? Such awareness may not be important as long as he gets where he wants to go, but can the same be said of solving mathematical problems?

Materials Needed

A teddy bear with jointed limbs.

Procedure

To first investigate the child's awareness of walking on all fours, ask him to walk on all fours a distance of about 10 meters and then to explain verbally how he did it. Next ask him to show how to do it with a teddy bear (with jointed limbs). If necessary, get on the floor with the child and ask him to indicate which limb to move first.

Levels of Performance

Stage 1A. Four-year-old children, or slow developers up to the age of 7, respond that they move everything at the same time or their hands and then their feet (the Z solution).

Stage 1B. This child, 5 to 6 years old, gives an N-type description: he moves the hand and foot on one side and then the hand and foot on the other.

Stage 2. Half the 7- to 8-year-olds and two-thirds of the 9- to 10-year-olds give the correct X solution: they alternate—left foot, right hand, right foot, left hand, for example. It is not until this stage that there is a clear grasp or cognizance of the individual movements.

Why is the child so late to become cognizant, since the necessary

---

[1]Jean Piaget, *The Grasp of Consciousness—Action and Concept in the Young Child* (Cambridge, Mass: Harvard University Press, 1976), pp. 1-11.

Three children having just crawled try to describe what they did. The 6-year-old says "I just crawled." The 7- and 9-year-olds, both Stage 1A, describe a Z solution (arms, then legs).

Mark, 6, shows an N solution to describe crawling (left arm and left leg moved simultaneously).

sensorimotor regulations or actions can be performed by a baby? The baby "knows how to" walk on all fours, but he does not "know" or is not conscious of the steps involved in the process.

The child at Stage 2 is able to "think about" what he did—to break down an automatic sequence of movements in terms of their correct order. This Piaget calls the beginning of operating reversibility. It includes a certain level of conceptualization to counterbalance what was before purely automatic. The Stage 2A child may have to "stop and think" as he crawls to answer correctly, while the Stage 2B child responds immediately and correctly. Even so, this cognitive ability comes long after knowing how to walk on all fours.

Teaching Implications

See Teaching Implications, Activities 5-2 through 5-4.

THROWING AN OBJECT WITH A STRING

Purpose of Activity

"Knowing how to" is not the same as "knowing," as demonstrated in the preceding activity. We may "know how to" multiply without being cognizant of what is involved in the process; that is, we may know how to put the correct numerals in the proper places based on certain rules. Similarly, we may know how to do the activity described below without understanding the process involved.

In Activity 5-1, children were found to become conscious of the order of their actions at 7 to 8 years of age. In this and subsequent activities, consciousness is found to come much later, at 11 to 12 years of age.

Materials Needed

1. A ball, 5 centimeters in diameter, attached to a string.
2. A rectangular box.

Procedure

Show the child the ball attached to the string. Hold the end of the string and make the ball swing in a circle in a plane just above the floor. Then ask the child where the ball will go if the string is let go. First make the ball swing clockwise and then counterclockwise to see if the child in the second instance predicts an opposite direction for the ball to go when released.

Next ask the child to try it himself. After this, place a rectangular box outside the circle of the swinging ball and ask the child to let go of the string so that the ball will go in the box.

To describe for the reader the release positions, we will use a clockface with numerals, as follows:

With the child outside the circle at the six o'clock position, the ball must be released at about nine o'clock to go in a box at twelve o'clock, if the motion of the ball is clockwise. If counterclockwise, the ball should be released at the three o'clock position.

Levels of Performance

It is interesting that the sensorimotor actions necessary for success in the task are performed long before the child is conscious of what happens to the ball in its flight.[2]

Stage 1A. The 5-year-olds may be successful in getting the ball to land in the box, but they cannot explain it:

"If you let it go, where does the ball go?"

"It goes everywhere."

"To get the ball in the box, when do you let it go?"

"When you see the box."

By trial and error, 5-year-olds learn when to release the ball to get it in the box. The child's cognizance of his actions places more emphasis on where he stands and the force he uses rather than on his observance of what the ball does. For example, the Stage 1A child believes that he must stand at the six o'clock position to make the ball go in a box at the twelve o'clock position.

The child knows that the ball leaves the circular orbit when the string is released, but he cannot describe its trajectory. He thinks that the ball always goes to the left when released from a counter rotation and to the right when released from a clockwise rotation.

Stage 1B. This child does not believe that he has to be opposite the box (at six o'clock to make the ball go in the box at twelve o'clock), but he is not cognizant of the release point, saying, for example, "You release at twelve o'clock to make the ball go in the box at the twelve o'clock position."

Stage 2A. At this stage, ages 8 to 9, children are successful at the sensorimotor level; they "know how to" regardless of where they stand or where the box is placed. They do not ask to be placed opposite the box (at six o'clock if box is at twelve o'clock). However, they are still not conscious of correct release points, insisting that the ball is released at twelve o'clock if the box is at the twelve o'clock position.

Stage 2B. At around 9 years of age, the beginning of cognizance or consciousness of what is going on occurs. At first the child re-

---

[2]Ibid., p. 14.

sponds as do the children at the lower stages—that to get the ball in the box at twelve o'clock you release at twelve o'clock (even though they release it correctly at nine o'clock for a counterclockwise motion of the ball). Then they quickly correct themselves.

The interviewer swings the ball and asks the child for instructions as to when to release the ball with the box at twelve o'clock. As the ball swings in a circle, Ped, age 9, says release at twelve o'clock, then nine o'clock, then twelve again. When the ball is released at twelve o'clock, Ped says, "You've got to let it go here [at ten o'clock]." Then trying it himself with the ball swinging counterclockwise, he releases at two o'clock:

"Where did you let it go?"

"Here [at seven o'clock]."

But when the interviewer starts to release at seven o'clock, Ped immediately says, "I bet it will go there [correct prediction of trajectory]." Subsequently, he correctly indicates his own release points; that is, nine o'clock for clockwise motion of the ball with the box at twelve o'clock and three o'clock for counterclockwise motion of the ball.[3]

Stage 3. Not until 11 to 12 years of age will the child be cognizant of the fact that where the ball goes depends on its release point and the direction of rotation. There is no need for trial and error to achieve success. A distinction can still be made between 3A and 3B, however. At 3A children still cannot correctly describe the trajectory of the ball after the string is released. Some say the path of the ball curves; others say it zigzags. At Stage 3B, however, the child describes the path correctly as tangential, even if he does not know the word:

"Where does the ball go?" Eri, 12, says, "I see in my head where the ball goes. It follows the movement of the circle a bit." But he draws a tangential path for the ball from the point where it is released.

"What are the angles like?"

"At the place where you let the ball go, it's perpendicular. There is a right angle." (Meaning a right angle between the radius of the circle to the ball at point of release and its tangential path therefrom.)[4]

---

[3]Ibid., pp. 28–29.
[4]Ibid., p. 34.

Teaching Implications

This activity provides clear evidence that knowing "how to" do something, such as how to get a correct answer by a sensorimotor process, is very different from "knowing" or being "conscious" or "cognizant" of the processes involved. Furthermore, the time lag is one of years—not months or days.

How important then this is if our teaching processes consider whether the child knows or is conscious of what he is doing or whether he is operating at the sensorimotor level—"knowing how to" get an answer without being conscious of the processes involved!

The activity calls for inferences not possible in children until the formal operational thought level—that is, listing possible factors involved in getting the ball in the box; ruling out those that are not responsible, such as where one stands or the force used in swinging the ball; and finding the point at which the ball must be released and the path it must follow in order for it to land in the box.

Piaget asks how this correction of release points comes about and concludes that it is the progress that generally takes place at the age when children first grasp the significance of direction and vectors in the construction of natural systems of coordinates.[5] (See also Activities 3-3 through 3-5.)

[5]Ibid., p. 30.

SPINNING A PING-PONG BALL OR HOOP

Purpose of Activity

In this activity, children carry out two simultaneous actions—push a ball or hoop forward and, at the same time, put a backspin on it to make the ball come back to them. Many of us have done this as children, but being able to or knowing how to do it cannot be equated with a consciousness of what is going on. There is a time lag of four or five years.

Materials Needed

A ping-pong ball or a hoop.

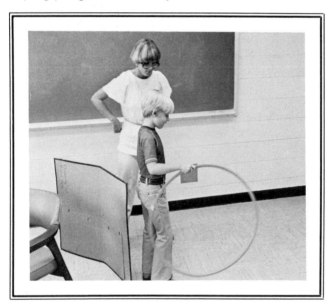

Scott, 7, is unable to put a reverse spin on the hoop after the teacher has done it with the screen hiding the motion of her hand.

Procedure

Ask the child if he thinks he can make a ping-pong ball go forward and then come backward without touching it again or letting the ball hit a wall. Then ask him to try it. If he is unsuccessful, demonstrate with your hand behind a screen, asking the child to watch the ball. If the child still cannot do "the same thing," show him the whole process and ask him to try it himself. At 5 or 6 years of age, the child is usually successful at least once during several attempts; he has constant success at 6 to 7 years of age.

As a next step, ask the child to describe the ball's movements with such questions as "What did the ball do?" "What did you do?" and "That's all?"

Levels of Performance

Stage 1A. This child is sometimes successful in performing the task, but when asked why the ball comes back says, "Because you throw it so hard" or "It goes straight and then it comes back."

Stage 1B. There is an awareness that the ball is spinning as it goes forward, but the child does not realize that it is a reverse spin. He is aware that his fingers are a cause: "You slide your fingers on the ball."

Stage 2A. At age 7 to 8 there is a partial understanding of reverse action. This child still thinks the ball is spinning forward, but he realizes that he is performing two actions—both pressing down on the back of the ball and making a withdrawing movement at the same time. He still thinks the ball is spinning forward, however.

"What makes it come back?"

"It gets enough push" or "I press on the ball and pull my hand backward."

"What makes it come back?"

"The bounce."[6]

Stage 2B. After many trials, errors, and contradictory statements, the 9- to 10-year old child is finally conscious of the reverse rotation of the ball.

Stage 3. At 11 to 12 years of age (with a few precocious children at 8 to 9), the child is immediately conscious of the fact that he is simultaneously carrying out two actions—propelling the ball forward and at the same time making it spin backward. He may first have to watch the interviewer do it, but he can analyze the action successfully.

---

[6]Ibid., p. 60.

Teaching Implications

The Stage 3 children, at the Formal Operations stage, can make inferences from what the ball "does" as a result of their own "actions." They coordinate the two categories of observation—what the ball does and what they do—and from this process, draw their inferences.

The same general results will be obtained with a large hoop as with a ping-pong ball. While it takes more strength and more skill to keep the hoop upright, the reverse rotation of the hoop is easier to see.

Again, as in the preceding activity, there is a time lag of approximately six years, from age 6 to age 12, in accomplishing the necessary sensorimotor regulations of "knowing how to" as contrasted to the formal operations of being conscious of or "knowing" by inference what is actually taking place. And again this implies careful consideration of just what it is that the teacher is truly teaching. Is it only how to do something? Performing is not understanding. Are the children aware of the processes involved?

BUILDING A ROAD
(SLOPE AND ANGLE)

Purpose of Activity

Activities 5-2 and 5-3 were concerned with the child's ability to analyze and explain what happened to a moving object. This activity involves the use of static or immobile material to explore the child's ability to select material and put it together properly to build a "road" to the top of a "mountain" (a cardboard box).

Materials Needed

1. Nine bricks, one thin one, a taller one, a cylindrical one with grooves, and six small cubes.
2. Small strips of wood, from 15 to 25 centimeters long. One strip should be 50 centimeters long.
3. A doll or small car.

Procedure

Place the doll or car 50 centimeters from the box. Show the child the materials and ask him to build a road to the top of the mountain (box) for the doll (or car) to reach the top.

Levels of Performance

Stage 1. To solve this problem, the child has to relate the distance of the doll from the box and the height of the box, which of course is a slope. He has to imagine the slope without seeing it (until a proper road is built). This, the Stage 1 child cannot do. The starting point for the doll is some distance from the box, which he forgets or ignores, building his tower next to the box. If reminded of the starting point, the doll, he builds a horizontal road to the foot of the box. Even if asked whether the long strip of wood would help, the Stage 1A child cannot envision its possible use in making a slope. He is unable to think of the blocks as intermediary points, or he forgets about them. The Stage 1B child realizes the possibility of slope, but he cannot use the long stick correctly. He places it almost vertically next to the box so that it reaches the summit.

Stuart, 6, Stage 1A, unable to use the idea of slope, builds his road horizontally to the base of the box and then vertically.

Stage 2. These children, 7 to 10 years of age, start off immediately to make a slope, but they may still make isolated towers—a Stage 1 characteristic.

The Stage 2 children do not just center their attention on the goal, the mountain top, but consider the starting point as well and attempt to make links between the two. At Stage 2A, however, children are unable to make a proper link-up—their strips are too short, there is no link between them, the slopes are too steep, and so forth. At Stage 2B, there is an anticipatory scheme. Mistakes are made, but they are minor and quickly corrected.

Stage 3. These children react in almost the same manner as do Stage 2B children. They do pay more attention to the accuracy of their solution—location of pillars, wedging bricks, weight of road in relation to strength of pillars, and so on. But Piaget remarks that this has no bearing on the processes leading to consciousness or cognizance of the solution to the problem; that is, the necessary retroactive and anticipatory mental regulations that the Stage 2B child also has.[7]

---

[7]Ibid., p. 118.

To review the three facets of the problem, the children have to

1. relate the horizontal ground at the starting point (doll) to the vertical mountain (box)—that is, find the solution of the slope;
2. discover the possible relationship between this slope and the objects to be used as intermediaries (blocks, sticks, etc.); and
3. determine the possible relationships among the objects themselves and decide if they have a direct link to the first two types of relationships.

Teaching Implications

Just what is it that teachers teach? Is the present trend "back to the basics" just knowing how to get answers using a certain set of rules?

A child of 6 has the necessary sensorimotor regulations to know how to respond or write "6" when the teacher shows him a set with six objects in it, but is he cognizant of the inclusion and reversibility relationships involved for understanding conservation of number? Even though he knows how to write "7" when he sees 4 + 3, does he know the inclusion relation involved? (See Activity 2-8.)

Similarly, the 10- to 11-year-old may know how to use the rule for proportions, $a/b = c/d$, but not be cognizant of the basic structures involved.

It is also worth pointing out that the problems considered in this section basically involve geometrical ideas (ball or sphere, circular swing, tangential paths, angles and slopes, and even walking on all fours on the floor). All actions in the physical world or space may be thought of as involving geometrical ideas.

6
Mathematical Memory

One of the traditional methods of teaching is "show and tell." This method is based on the supposition that memory is perceptual in character—that what is seen or heard is a true copy of reality. But the intellectual processes available to the child must be considered when logical structures are to be taught.

Piaget's recent book, *Memory and Intelligence,* relates the idea of various mathematical processes to children's ability to remember them. Can the child recall or remember, for example, a set of sticks that has been shown to him arranged in a certain pattern; e.g., ordered by height from shortest to longest? According to Piaget, children cannot remember what they have seen if a logical or mathematical process such as ordering is involved until they have reached a certain developmental level. Memory, then, pertains to the ability of the mind to reconstruct that to which it has been exposed. A distortion will result until the child has the necessary cognitive processes.

This section provides elegant proof of the existence of developmental stages in mental maturation. More than that, it points out a basic faulty teaching method—"show and tell."

Children may be "taught" knowledge of a sensorimotor sort, as in learning to name colors. Perceptual cues provide the right answer. Children may be "shown" a green object, for example, and "told" that it is green. They can then identify other objects as "green" if they are not color blind. However, such show-and-tell procedures will not work in teaching logical structures such as addition, transitivity, inclusion, the commutative property, ordering, and measurement. In short, all the various logical or mathematical structures cannot be taught until the child has the comparable psychological structures.

Memory of such mathematical structures will be studied in terms of three activities:

6-1. Memory and Ordering

6-2. Memory and the Concept of Horizontal

6-3. Memory, Conservation of Number, and Conservation of Length

MEMORY AND ORDERING

Purpose of Activity

The purpose is to demonstrate that the memory of a set of sticks already ordered by height and "shown" to a child cannot help her build the series correctly at a later date if she does not yet have the necessary logical structures.

Materials Needed

1. Ten sticks of different size, ranging from 9 to 16 centimeters in length.
2. Paper and a pencil.

Procedure

Show the child the sticks arranged in order from shortest to longest. Then ask her to describe the series, to take a good look at it and remember what she has seen.

About a week later ask the child to:

1. trace out with her finger on a table what she saw;
2. draw a picture of it (direct copy drawings are not made successfully until about 5 years of age);
3. use the sticks, themselves, and reconstruct the ordered set (to determine her operational or intellectual level);
4. draw the series of sticks once again as a test of her ability to make a direct reproduction (if the drawing, Item 2, is incorrect);
5. describe verbally the layout of the sticks when asked such questions as "What is this?" "Are they all the same?" "How do they differ?"
6. alter her verbal description (Item 5) when asked to begin with a particular stick

181

After one week there is likely to be a fairly marked correspondence between intellectual or operational level and the organization of the memory. The 4- to 7-year-old draws not what she saw but what her intellectual structures allow her to draw, which is incorrect.

If possible, the child should be seen six to eight months later and asked if she can remember what was done in order to see if her memory has, in fact, developed.

Levels of Performance

Piaget found, in testing twenty-four children eight months later, that only two made no progress. Twenty-two produced memory drawings representing a marked advance from the old, "thus giving striking proof that their memory had evolved in the course of these eight months [as well as intelligence]."[1]

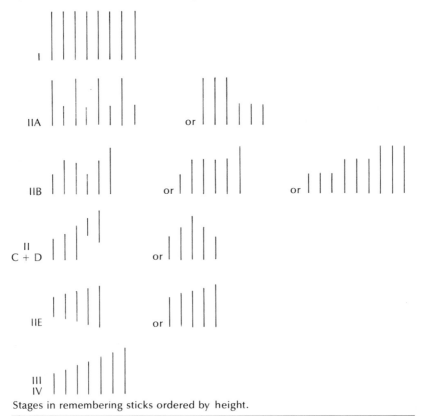

Stages in remembering sticks ordered by height.

---

[1]Jean Piaget and Barbel Inhelder, *Memory and Intelligence* (New York: Basic Books, Inc., 1973), p. 41.

Two children (one 4- and one 5½-year-old) progressed from Stage 1 in the test after one week to Stage 2A eight months later. Two more 4-year-olds and a 6-year-old advanced from 2B to 2E in their ability to remember during the eight-month period. In the first testing after one week they drew a dichotomy—4 long and 4 short sticks. In the testing eight months later they drew a fan shape (sticks getting longer at both ends) or a trichotomy of 3 sticks with the 2 succeeding sticks, each longer than the one before.

One bright 4-year-old child produced a completely correct series of 10 sticks ordered by length during the test eight months later. She had moved during this eight-month period from a Stage 2B level to the operational level of Stage 3.

To remember a series of 10 sticks—a, b, c, . . .—arranged in order by length from shortest to longest, the child must realize the relation not only of b to a and c to b, but also of c to a. If b is longer than a and c is longer than b, then c is longer than a (the construct of transitivity). Also necessary for correct ordering is the understanding of the dual relation for each element—that it is longer than the preceding element and shorter than the one to follow.

Teaching Implications

This and the next two activities provide striking evidence that "show-and-tell" teaching procedures do not teach. The mind does not work like a camera. Children can only remember what they can reconstruct of what they see or hear. Various stages of development determine what a child can learn or remember. (See also Ordering and Seriation, Activity 2-2.)

MEMORY AND THE CONCEPT OF HORIZONTAL

Purpose of Activity

To organize the relation between objects in space in terms of distance and direction requires spatial concepts such as horizontal and vertical or north-south and east-west. These concepts are not abstracted by children until around 9 years of age (see Activities 3-3 and 3-4).

Materials Needed

1. A jar partially filled with liquid.
2. A table.

Procedure

Show the child the jar. Ask her what the water in the jar will look like if the jar is tilted "like this." (Tilt the jar at an angle of 45 degrees while hiding the water surface.) Then tilt the jar with the child watching and ask her to look and remember what she saw.

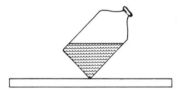

Sixty-six 5- to 9-year-olds were divided into two groups to determine to what extent memory is a factor in such a task. After being shown the bottle, one group was seen one hour later and again one week later. The second group was seen again one week later only. When the children were seen again they were asked to draw the bottle and liquid as they remembered it.

Levels of Performance

In both groups, correct responses were given by only one-fourth of the 5- to 7-year-olds and by only one-third of the 8-year-olds. There was no significant difference in the results in retesting one hour after and one week after the original test.

Six months later, the test was again conducted with 55 of the original 66 children. All of them remembered a partially filled bottle. The percentage of correct responses among the 7- to 9-year-olds during this six-month period increased from 27 percent to 50 percent. This result suggests partial progress and improvement in memory based on new cognitive processes developing during the six months between the tests.[2]

The responses could be fitted into the following mnemonic types:

Type I: The bottle is drawn in an upright or inclined position, with the liquid covering one of its walls.

Type II: The bottle is drawn in an inclined position, but the surface of the liquid is drawn parallel to the bottom; i.e., at an angle of 90 degrees to the axis of the bottle.

Type III: The bottle is inclined, and so is the surface of the liquid.

Type IV: The bottle is upright, but the surface of the liquid is inclined at an angle of 30 to 40 degrees.

Type V: The problem is suppressed, either because the bottle itself is drawn in a horizontal position or else because it is inclined or erect, but full.

Type VI: Correct remembrance.[3]

Teaching Implications

See Teaching Implications, Activities 6-1 and 6-3.

---

[2]Ibid., p. 302.
[3]Ibid., p. 300.

MEMORY, CONSERVATION OF NUMBER, AND CONSERVATION OF LENGTH

Purpose of Activity

Children achieve the concept of conservation of number before that of conservation of length. So what happens in the interim, when they attempt to remember an illustration involving both concepts?

Materials Needed

An illustration, such as that used by Piaget:

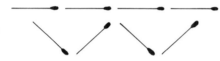

Procedure

Children who understand conservation of number realize that the two rows have the same number of matchsticks. However, if these children do not understand conservation of length, they think the top path is longer since the end points are further apart.

What effect, then, does memory have on attempting to reconstruct this illustration at a later time? Do intellectual processes attempt to resolve this confiict for the child who is operational for the conservation of number concept and preoperational for the conservation of length concept? This is, in fact, the case.

Levels of Performance

Children who do not understand conservation of length resolve the conflict by drawing more matches in the bottom row to make the two rows the same length:

or, rather ingeniously, lengthen the matches in the second row:

186

Thus these children draw an illustration that is different from the one they saw previously using the intellectual processes available to them in an attempt to resolve the cognitive conflict. They do not necessarily "see" or "remember" or "copy" what is shown, but rather *reconstruct* what is shown, particularly if there are logical structures involved in the problem, as is usually the case in mathematics.

Teaching Implications

Piaget rejects the common notion of memory as a "copy" of reality. There does not even need to be an internal image for the first level of memory, that of **recognition.** Recognition is a part of every sensorimotor habit. The baby "recognizes" his mother or the feeding bottle and starts to suck without being able to construct a mental image of mother.

The second and third levels do involve the common notion of memory as the ability of "recall," but to recall what? The second level of memory is an imitative one based on sensory impressions or perceptual data. This level Piaget calls **figurative knowing.** It involves recall, but the recall is often a distortion due to the intellectual level of the child.

Advocates of "learning theory" often contend that mental development, including the evolution of thought processes, is the result of learning that is largely externally directed or imposed, and that memory is nothing other than conservation of these results. These activities demonstrate that this is not the case. (See also page 5.)

The teacher must find ways of teaching other than "show and tell" in order for children to remember mathematical ideas. The students must be treated as active learners who construct their own reality—which, in fact, is what they do as just described with the matchsticks. Their success will depend on the cognitive processes available to them rather than on what they have been shown or told.

Memory does interact with intelligence, and sometimes there is spectacular progress in recall after a lapse of several months.[4] Is this

---

[4]Ibid., p. xi.

why the dawn seems to break suddenly for many students in mathematics?

The reader may be interested in other mathematical concepts related to memory, as described by Piaget in *Memory and Intelligence*. These include equivalency of sets, transitive relations, associative relations, serial correspondence, double classification, and intersection.

7
Chance and Probability

Many of the "why" questions children ask cannot be explained rationally. Sometimes events happen by chance—for no logical reason at all. But the children see a hidden cause.

For example, why is this stick longer than that one? Why does Lake Geneva not go all the way to Berne? Why do your ears stick out? These famous "why" questions ask for a reason where a reason exists but also where it does not—where the phenomenon happens by chance.

The Origin of the Idea of Chance in Children (Piaget, 1975) describes in detail the development of the concepts of chance and probability. Young children do not even consider chance as a cause of events. A fully developed formulation of the mathematics of probability is not realized until 12 to 15 years of age. However, appropriate activities involving these concepts may be begun in the elementary school years.

It is interesting to note that primitive man probably had no concept of chance. He saw every event as the result of visible or hidden causes and lacked the rational or experimental criteria to rule out even the strangest phenomena as happening by chance.

If the intuition of probability is not innate and not primitive, then how do these notions develop in the minds of children? Further, once there is a realization of events occurring by chance, how is progress made in quantifying probabilities? Not until 14 to 15 years of age will permutations be understood. What are the implications for the school curriculum?

Recent elementary school textbooks in mathematics do include units on probability in the upper elementary grades. The research and activities described in this section may help determine more specifically at what level such activities are appropriate.

Activities included in this section are as follows:

7-1. Events Occurring by Chance (Random Mixture)

7-2. Events Occurring by Chance (Tossing Counters)

7-3. Beginning to Quantify Probability (Random Drawing)

7-4. Quantification of Probabilities (Use of Proportions)

7-5. Combinatoric Operations (Combinations)

7-6. Permutations (Arrangements)

EVENTS OCCURRING BY CHANCE (RANDOM MIXTURE)

Purpose of Activity

If a box is tilted so that the objects within are mixed up in a random fashion, does the child realize that in the mixing process the objects move independently of each other (except for collisions)? Also, can the mixing process be reversed?

Materials Needed

1. A box with a divider.
2. Eight red balls and eight white.

Procedure

Ask the child what he thinks will happen to the balls if the box is tilted (tip the box but hold the marbles in place). What will be the arrangement of the balls when the box is tilted back to its original position? Will the red ones stay on one side or will they get mixed up and, if so, in what proportion?

Then slowly tilt the box to the right and then back to the original position, letting the child observe what happens—several balls are now in different positions. Ask him what will happen if we do it again. He finds after the second tilt that more balls are now in different positions. Ask him to predict what will happen after many tilts.

This procedure leads to an examination of the child's concept of permutations (see Activity 7-6).

Levels of Performance

Stage 1. From 4 to 7 years of age, the child realizes that there is a total displacement of the objects from their original order, but he often predicts they will ultimately return to their original positions.

191

A Stage 1 child, Vei (5 years 6 months), is asked, "What will happen to the balls if I tilt the box?"

"They will stay just as they are now."

The box is tilted and one red ball goes to the white ball side and one white ball to the red ball side. Vei then says they cannot come back the same.

"And if we do it again?"

"Then they will get more mixed up."

"Why?"

"Because two will roll to the other side."

"And if we continue all afternoon?"

"It will all come back in place as it should."[1]

This is a typical response demonstrating the failure to understand the irreversibility of random mixing.

Stage 2. Beginning about 7 years of age, the intuition of chance appears (along with the developmental stage of concrete operations).

A 7-year-old looking at the marbles in the tilted balance says, "It gets mixed up in all sorts of ways and then it comes back together . . . little by little it'll slowly get unmixed up."[2]

At Stage 2, mixture is accepted as a positive fact and is no longer, as in Stage 1, an unnatural state. But even though mixture is admitted, it is not understood.

Stage 3. Beginning at 11 to 12 years of age, the formal operational level, the process of random mixture is understood: that it is a process that is not reversible, that the more the board is tilted the more the marbles get mixed up, that the chance of the marbles returning to their original positions is very remote.

Teaching Implications

Children in elementary school are not ready for a study of random mixture until they reach the age of 11 or 12. (See also Teaching Implications, Activity 7-4.)

---

[1]Jean Piaget and Barbel Inhelder, *The Origin of the Idea of Chance in Children* (New York: W. W. Norton & Company, Inc., 1975), p. 7.
[2]Ibid., p. 19.

EVENTS OCCURRING BY CHANCE (TOSSING COUNTERS)

Purpose of Activity

The purpose of this activity is to determine children's concepts of events that occur by chance.

Materials Needed

Counters or coins, with a cross marked on one side and a circle on the other.

Procedure

Show the child the counters or coins and ask him which will turn up when the counters are tossed—a cross or a circle:

Dexter, 7, thinks more crosses will turn up again on the next throw.

"If I toss this counter, will we see a cross or a circle?"

"A cross," responds a 5-year-old.

"Why?"

"Because it's bigger."

The child tosses the coin and a circle turns up.

"How about if we do it again?"

"Cross."

"Now if we throw them all together [10 to 20 counters], what happens?"

"Don't know, we will have to see."

The counters are tossed and more circles than crosses turn up.

"Why are there more circles?"

"Don't know."

"What if we threw them again?"

"Maybe more crosses, since last time there were more circles."

"Could they all be crosses?"

"Yes, if we want them to."[3]

This experiment is then varied by substituting a set of false counters with the same mark (either a circle or a cross) on both sides, but the child is not allowed to see the substitution. The objective is to see how the child reacts to all the counters turning up the same way. And then he is asked if there might be a trick involved.

Levels of Performance

Stage 1.   In these children there is a certain docility of response. "They just fell that way" or "If I haven't seen it, I can't know."

Most of these children do not think any trick is involved or, if it is a trick, it is the power of the person—a trick you do with your hands.

Stage 2.   Usually around 7 years of age children refuse to accept the result with the false counters (same mark on both sides). The children do not believe the counters could all turn up the same way without some trick. They have a global sense of probability which denies the result. They realize the material itself must be "fixed" and check the other side of a counter to see. For some, checking one counter is sufficient to make the generalization that the counters must all be the same on both sides.

Stage 3.   At 11 to 13 years of age, there is more than an intuitive or global sense of probability—there is a beginning of quantification. A

---

[3]Ibid., p. 99.

12-year-old when asked if there will be more crosses or more circles if 20 counters are thrown responds, "It's chance, it depends." And when asked, "Is it more likely to get ten circles and ten crosses with one thousand or ten thousand thrown?" he responds, "With a million."[4]

Teaching Implications

Children are not ready to quantify probability by tossing counters until the formal operational level—11 or 12 years of age. (See also Teaching Implications, Activity 7-4.)

---

[4]Ibid., p. 106.

BEGINNING TO QUANTIFY PROBABILITY (RANDOM DRAWING)

Purpose of Activity

The purpose is to investigate how children solve a problem of probability involving random drawing.

Materials Needed

Two identical sets of different colored counters, with a different number of counters in each color.

Procedure

Leave one set of counters on the table in order for the child to see how many of each color there are. Mix up a duplicate set in a bag

Scott, 7, predicts he will pick a red one from a set of 6 red, 4 blue, and 1 white; but he cannot give a reason.

and ask the child to reach in and pull out two counters. Before he looks at the counters, ask him to predict what color they will be.

Using 6 red, 4 blue, and 1 white, a 6-year-old predicts that the first counter picked will be white:

"Why?"

"Because there is only 1 white one."

The white one is then taken out of the sack while the child watches and he is asked to predict what color the next one picked will be:

"Blue."

"Why?"

He opens his hand, finds a red one, and asks, "Why is it red?"

Levels of Performance

Stage 1. As described in the preceding, there is no conception of chance or probability. The difference in quantity of the various colors does not seem to be a factor in guessing what color will be picked. The prediction may be red:

"Why red?"

"Because I like red a lot."

Stage 2. Here there is a beginning of quantification, but as the counters are taken out, the children make predictions as if all of the original counters were still in the bag. They are unable to make the adjustment after each trial for the number of counters still in play.

Stage 3. At 11 to 12 years of age, children do take account of counters that are removed as subsequent predictions are made. A 12-year-old picking from a collection of 14 green, 10 red, 7 yellow, and 1 blue predicts that the first two will be greens. But he gets a red and a yellow:

"And now, how about the next pick?"

"Two greens." (And he gets 2 greens.)

"And for the next pick?"

"There is still a better chance for 2 greens since 12 of them remain."[5]

At Stage 3 there is a refinement of the idea of chance, and the calculation of probability is based on the totality of possible combinations. The notions of chance and probability are by nature essentially combinatoric.[6]

---

[5]Ibid., p. 127.
[6]Ibid., p. 128.

Teaching Implications

Children are not ready to quantify probability by random drawing until 11 or 12 years of age—the formal operational stage. (See also Teaching Implications, Activity 7-4.)

QUANTIFICATION OF PROBABILITIES (USE OF PROPORTIONS)

Purpose of Activity

Piaget asks why children are so late in recognizing the numerical relations involved in understanding probabilities. Also, what logical and arithmetical operations are necessary to perform such tasks successfully?

Materials Needed

Two small sets of white counters, some of which have a cross on the back.

Procedure

Show the child two sets of counters, allowing him to look at the back side of the counters in each set. Then move the counters around within their respective sets so that the child does not know which ones have crosses on the back. The question then is, "From which set would it be easier to pick a counter with a cross on the back?"

Levels of Performance

Stage 1. At 4 to 7 years of age, children ignore the quantitative relations involved. Having seen that in one set one out of three has a cross on the back while in the other set all three have a cross on the back, a Stage 1 child, asked to pick one with a cross on the back, still picks the set with only one having a cross on the back:

"Why did you pick that set?"

"It's easiest because there is only one cross."

In another problem in which one set of two counters has one with a cross on the back and a second set of four counters has two with crosses on the back, the Stage 1 child picks the second set because "There are two [counters with crosses]."

Stage 2. At about 7 years of age, when the first logical arithmetical operations are acquired, children can solve problems involving only a single variable. For example, if both sets contain three counters, one set having a single counter with a cross on the back and the

199

other set having two counters with crosses on the back, the children realize the second set involves a 2/3 chance as opposed to a 1/3 chance in the first set.

Similarly, if the numerators are the same and the denominators different—2/3 and 2/4, for example—these children realize the chance is better to pick a cross from the 2/3 relationship.

Stage 2 children, like Stage 1 children, are still unable to compare two variables—fractions such as 1/2 and 2/4 in which both numerators and denominators vary. A 10-year-old says, "Here [1/2], no, there [2/4]."

"Then is one more certain?"

"No, they are the same, but we are more certain here [2/4] because there are two with crosses [very perplexed]."[7]

Stage 3. At the formal operational level, children can solve problems of proportionality. With the two fractions 1/3 and 2/5, the child picks the 2-to-5 relationship realizing 1/3 is the same as 2/6, which means 2/5 is a better chance. Similarly, with 1/3 and 8/17, a 12-year-old responds, "It's easier with 8/17. If it were the same, it would be 8/24."[8]

Teaching Implications

Piaget concludes that fundamental probabilistic notions do not develop until the formal operations level—around 11 to 12 years of age. The reason is that, psychologically, formal operations are operations of the **second power,** or operations requiring previously learned operations; that is, concrete operations. Formal operations are more abstract and require a hypothetical-deductive power of considering the possible, as well as probable, and the different logical connections, such as disjunctions, before probability can be quantified.[9]

It would follow, based on these conclusions, that probability topics in the elementary grades are out of place, except for an incidental or intuitive approach or for gifted children.

---

[7]Ibid., p. 154.
[8]Ibid., p. 159.
[9]Ibid., pp. 159–60.

COMBINATORIC OPERATIONS (COMBINATIONS)

Purpose of Activity

In Activity 7-3, combinations were considered in terms of picking two counters from a bag and predicting what color they would be. In this section, children are asked to combine different colors in all possible ways.

Materials Needed

1. Three sets of counters—white, red, and green.
2. Counters of various colors (see Levels of Performance, Stages 2 and 3).

Procedure

First show the children counters in three sets by color—white (w), red (r), and green (g). Tell them that the counters are little boys and ask how they can leave two at a time.

Levels of Performance

Stage 1. Children 6 to 7 years of age have no real sense of combinations. A 6-year-old puts two whites together, two reds together, and two greens together:
"What else?"
"This" (meaning another pair of whites).
"Didn't you do that already?"
"Yes."
He then puts a red and green together, then white and green, and then another red and green.
"Didn't you do that already?"
He takes away the second red and green.
"Is that all?"
"Yes, we have them all. We need another color."[10]

---

[10]Ibid., p. 165.

This child misses one of the three possible combinations of different colors—red and white.

Stage 2. At 7 to 11 years of age there is a search for a system and a beginning of quantifications, but the process is trial and error.[11] Given four different colored piles of counters (A, B, C, D), an 8-year-old pairs them as AB, CD, then AC, AD, CD, then removes the second CD pair saying, "That's all."

"How can you be sure?"

"Oh, still this one [BC]."

Piaget concludes that the general characteristic of this stage (and the prescientific mind as well) is to start using an additive idea rather than intersection or multiplicative associations.[12]

Stage 3. At 11 to 12 years of age, the formal operational level, some children discover a system in which no pairing is skipped. This is in contrast to Stage 2 where the system is incomplete and has to be finished with trial and error or empirical searching.

An 11-year-old considering six piles of counters, each pile a different color, immediately finds a system for associating the color in Pile A with all the rest—AB, AC, AD, AE, AF, then B with the following colors—BC, BD, BE, BF, and then C in the same way—CD, CE, CF. Then DE, DF, and finally EF.[13]

Teaching Implications

The activity demonstrates that combinatorial operations are not developed fully until the formal operational level (hypothetical-deductive thought), around 11 or 12 years of age. Children's ideas of chance and probability depend in a very strict manner on the evolution of combinatoric operations.[14]

What makes combinations more difficult than simple seriation or correspondence? "The answer is apparently that these correspondences are not independent of each other in the case of combinations, but that they constitute a unique system of such a sort that

---

[11]Ibid., p. 167.

[12]Ibid., p. 169.

[13]The formula for the combination of n things taken 2 at a time is $n(n-1)/2$. Pairing colors where n or the number of colors is 6, the possible combinations are

$$\frac{6(6-1)}{2} \quad \text{or} \quad \frac{6 \cdot 5}{2} \quad \text{or} \quad 15$$

[14]Ibid., p. 161.

what is done first determines what follows."[15] Formal operations are operations of the second power; that is, operations bearing on other operations.

---

[15]Ibid., p. 172.

PERMUTATIONS (ARRANGEMENTS)

Purpose of Activity and Discussion

The purpose is to explore children's understanding of the various orders in which a set of objects can be arranged. **Permutations** are the number of different orders in which a set of elements can be arranged if all elements are used in each arrangement. For example, how many ways can you seat two people in two chairs?

For only two people, A and B, there are only two possible orders or permutations: they can be seated in order AB or order BA. Adding a third person, C, how many ways can the 3 be seated? The third person, C, can be seated in back of, between, or in front of the pair AB:

ABC

ACB

CAB

He can also be placed in back of, between, or in front of the pair BA:

BAC

BCA

CBA

This produces six possible permutations for three elements.

For four elements—A, B, C, D—we can put D with the triplet ABC in four more different ways (before, after, between A and B, and between B and C):

$DABC$

$ABCD$

$ADBC$

$ABDC$

Adding a fourth element allows four more permutations for each of six triples, thus producing 4×6, or 24 permutations for four elements.

Sharon, 7, finds 4 of the possible 6 permutations for three objects. Darin, 8, finds 5 of the possible 6.

Organizing this information for possible permutations of two, three, and four elements:

Number of elements	Permutations
2	2
3	6
4	24

Children cannot be expected to discover the formula for permutations, but they can develop a system such as that described above involving a small number of objects. The formula for the permutation of three elements is called 3 factorial and is written as 3!. It represents the product of the number of the elements (3) and all the counting numbers less than that number. Hence,

$3! = 3 \cdot 2 \cdot 1 = 6$

For four elements:

$4! = 4 \cdot 3 \cdot 2 \cdot 1 = 24$

For five elements:

$5! = 5 \cdot 4 \cdot 3 \cdot 2 \cdot 1 = 120$

For n elements:

$n! = n(n-1)(n-2) \cdots 3 \cdot 2 \cdot 1$

The psychogenesis of permutations does not usually occur in children until they reach 14 to 15 years of age. This is an interesting contrast to the development of combinations which occurs two to three years earlier, at 11 to 12 years of age.

Materials Needed

Cardboard shapes or counters, each a different color, to represent little men.

Procedure

To begin a study of permutations, ask the children to look at two "little men," A and B. The men are going to take a walk. "Can they be placed in a different way to take a walk? Like this [BA]?"

A 6-year-old, Stage 1, answers yes.

"And if we add another little man [displayed as ABC], are there still other ways?"

The child arranges the men as BAC and then as CBA. He then says that he thinks that is all.

Levels of Performance

Stage 1. At this level there is no system of analysis.

Stage 2. Here, partial systems are developed. The 8- to 10-year-olds (Stage 2A) are successful with three elements finding all 6 permutations, but are unable to continue with the same process for four elements.

At Stage 2B, children generalize what they did with three elements and apply it to the problem of four elements, even though they are unable to solve the problem completely, finding fewer than 24 possibilities for four elements.

Stage 3. At 14 to 15 years of age, a complete system is developed.

A 14-year-old in considering three elements says, "Oh, I see already that I can put each one twice in the first position. That will make 6."

"What about four elements?"

"I can put the new color first, and I can put it there 6 times. Then I can use the same system for each color. That will be 24 possibilities."

"For five elements?"

"There are 120 ways."

"Why?"

"If there are two colors, then we have 1 times 2. If there are three, we have 1 times 2 times 3. If there are four, we have 1 times 2 times 3 times 4 . . . ."[16]

This youngster has even developed the formula.

As to why the discovery of permutations is later than that for combinations:

> The reason . . . no doubt is that combinations consist simply in associations effected according to all the possibilities, while permutations, which are much more numerous, imply an ability to relate according to a mobile system of reference . . . . While the operations of combinations constitute a simple generalization of those of multiplication, permutations furnish the true prototype of relations or operations bearing on other operations.[17]

Stage 3, therefore, is divided into two substages—Stage 3A, marking the beginning of formal thought, and Stage 3B, as the level of equilibrium of this mode of structuring.

Teaching Implications

If the ideas of random mixture and probability are not understood until the formal operational level—around 11 or 12 years of age, or even later as the activities in this section indicate—then another look should be taken at the present trend toward making probability a part of the elementary school math experience. If probability is introduced, it should be done in a very limited way on a concrete level using objects that can be seen and handled.

Probability activities may serve to enrich the curriculum for

---

[16]Ibid., pp. 191–92.
[17]Ibid., p. 194.

brighter children. Teachers seem often to be at a loss for activities for gifted children; these may provide one answer.

Piaget describes three major stages of development related to understanding chance and probability. The first is before 7 or 8 years of age, characterized by the absence of elementary logical and arithmetical operations that involve reversibility (grouping of classes and relationships of whole and fractional numbers). Reasoning at this level is prelogical.

From 7 or 8 to 11 or 12, there is a gradual construction of groupings of logical order and by numerical sets—but on a concrete plane. Children must be allowed to see and handle the objects involved and to physically explore possible groupings or arrangements.

Not until 11 or 12 years of age, the stage of formal thought, can several systems of concrete operations be tied together at the same time and then translated into their hypothetical-deductive implications; that is, into terms of the logic of their propositions.[18] For permutations, readiness is even later—14 to 15 years of age.

---

[18]Ibid., p. 213.

Index

DATE DUE

SUGGESTED ACTIVITIES FOR EACH AGE

Ages	Logical classification	Number	Space orientation	Measurement	Knowing versus performing	Mathematical memory	Chance and probability
4	1-1, 1-2, 1-3, 1-7, 1-8	2-1, 2-2, 2-3, 2-4, 2-5, 2-6, 2-13, 2-15	3-1, 3-2, 3-4	4-4, 4-12		6-1	7-1
5	1-1, 1-2, 1-3, 1-4, 1-5, 1-7, 1-8	2-1, 2-2, 2-3, 2-4, 2-5, 2-6, 2-7, 2-8, 2-9, 2-10, 2-13, 2-14	3-1, 3-2, 3-4	4-1, 4-2, 4-4, 4-8, 4-12, 4-13, 4-14	5-1, 5-2, 5-3, 5-4	6-1, 6-2	7-1, 7-2, 7-3, 7-4
6	1-1, 1-2, 1-3, 1-4, 1-5, 1-6, 1-7, 1-8	2-1, 2-2, 2-3, 2-4, 2-5, 2-6, 2-7, 2-8, 2-9, 2-10, 2-11, 2-13, 2-14	3-1, 3-2, 3-3, 3-4, 3-5, 3-6, 3-7, 3-8	4-1, 4-2, 4-3, 4-4, 4-5, 4-6, 4-7, 4-8, 4-9, 4-11, 4-12, 4-13, 4-14, 4-15, 4-16	5-1, 5-2, 5-3, 5-4	6-1, 6-2, 6-3	7-1, 7-2, 7-3, 7-4, 7-5, 7-6
7	1-1, 1-2, 1-3, 1-4, 1-5, 1-6, 1-7, 1-8	2-1, 2-2, 2-3, 2-4, 2-5, 2-6, 2-7, 2-8, 2-9, 2-10, 2-11, 2-12, 2-13, 2-14, 2-15	3-1, 3-2, 3-3, 3-4, 3-5, 3-6, 3-7, 3-8	4-1, 4-2, 4-3, 4-4, 4-5, 4-6, 4-7, 4-8, 4-9, 4-10, 4-11, 4-12, 4-13, 4-14, 4-15, 4-16	5-1, 5-2, 5-3, 5-4	6-1, 6-2, 6-3	7-1, 7-2, 7-3, 7-4, 7-5, 7-6